Project-Based Inquiry Science™

GENETICS

Janet L. Kolodner

Joseph S. Krajcik

Daniel C. Edelson

Brian J. Reiser

Mary L. Starr

NSF

IT's ABOUT TIME®

HERFF JONES EDUCATION DIVISION

IT's ABOUT TIME®

HERFF JONES EDUCATION DIVISION

84 Business Park Drive, Armonk, NY 10504
Phone (914) 273-2233 Fax (914) 273-2227
www.its-about-time.com

Program Components

Student Edition	**Durable Equipment Kit**
Teacher's Planning Guide	**Consumable Equipment Kit**
Teacher's Resources Guide	**Multimedia**
	— **Genetics DVD**
	— **Netlogo Software CD**

Printed and bound in the United States of America.

ISBN-13: 978-1-58591-623-8

1 2 3 4 5 VH 12 11 10 09 08

This project was supported, in part, by the **National Science Foundation**
under grant nos. 0137807, 0527341, 0639978.
Opinions expressed are those of the authors and not necessarily
those of the National Science Foundation.

Principal Investigators

Janet L. Kolodner is a Regents' Professor in the School of Interactive Computing in the Georgia Institute of Technology's College of Computing. Since 1978, her research has focused on learning from experience, both in computers and in people. She pioneered the Artificial Intelligence method called *case-based reasoning*, providing a way for computers to solve new problems based on their past experiences. Her book, *Case-Based Reasoning*, synthesizes work across the case-based reasoning research community from its inception to 1993.

Since 1994, Dr. Kolodner has focused on the applications and implications of case-based reasoning for education. In her approach to science education, called Learning by Design™ (LBD), students learn science while pursuing design challenges. Dr. Kolodner has investigated how to create a culture of collaboration and rigorous science talk in classrooms, how to use a project challenge to promote focus on science content, and how students learn and develop when classrooms function as learning communities. Currently, Dr. Kolodner is investigating how to help young people come to think of themselves as scientific reasoners. Dr. Kolodner's research results have been widely published, including in *Cognitive Science, Design Studies,* and the *Journal of the Learning Sciences.*

Dr. Kolodner was founding Director of Georgia Tech's EduTech Institute, served as coordinator of Georgia Tech's Cognitive Science program for many years, and is founding Editor in Chief of the *Journal of the Learning Sciences*. She is a founder of the International Society for the Learning Sciences, and she served as its first Executive Officer. She is a fellow of the American Association of Artificial Intelligence.

Joseph S. Krajcik is a Professor of Science Education and Associate Dean for Research in the School of Education at the University of Michigan. He works with teachers in science classrooms to bring about sustained change by creating classroom environments in which students find solutions to important intellectual questions that subsume essential curriculum standards and use learning technologies as productivity tools. He seeks to discover what students learn in such environments, as well as to explore and find solutions to challenges that teachers face in enacting such complex instruction. Dr. Krajcik has authored and co-authored over 100 manuscripts and makes frequent presentations at international, national, and regional conferences that focus on his research, as well as presentations that translate research findings into classroom practice. He is a fellow of the American Association for the Advancement of Science and served as president of the National Association for Research in Science Teaching. Dr. Krajcik co-directs the Center for Highly Interactive Classrooms, Curriculum and Computing in Education at the University of Michigan and is a co-principal investigator in the Center for Curriculum Materials in Science and The National Center for Learning and Teaching Nanoscale Science and Engineering. In 2002, Dr. Krajcik was honored to receive a Guest Professorship from Beijing Normal University in Beijing, China. In winter 2005, he was the Weston Visiting Professor of Science Education at the Weizmann Institute of Science in Rehovot, Israel.

Daniel C. Edelson is Vice President for Education and Children's Programs at the National Geographic Society. Previously, he was the director of the Geographic Data in Education (GEODE) Initiative at Northwestern University, where he led the development of Planetary Forecaster and Earth Systems and Processes. Since 1992, Dr. Edelson has directed a series of projects exploring the use of technology as a catalyst for reform in science education and has led the development of a number of software environments for education. These include My World GIS, a geographic information system for inquiry-based learning, and WorldWatcher, a data visualization and analysis system for gridded geographic data. Dr. Edelson is the author of the high school environmental science text, Investigations in *Environmental Science: A Case-Based Approach to the Study of Environmental Systems*. His research has been widely published, including in the *Journal of the Learning Sciences*, the *Journal of Research on Science Teaching*, *Science Educator*, and *Science Teacher*.

Brian J. Reiser is a Professor of Learning Sciences in the School of Education and Social Policy at Northwestern University. Professor Reiser served as chair of Northwestern's Learning Sciences Ph.D. program from 1993, shortly after its inception, until 2001. His research focuses on the design and enactment of learning environments that support students' inquiry in science, including both science curriculum materials and scaffolded software tools. His research investigates the design of learning environments that scaffold scientific practices, including investigation, argumentation, and explanation; design principles for technology-infused curricula that engage students in inquiry projects; and the teaching practices that support student inquiry. Professor Reiser also directed BGuILE (Biology Guided Inquiry Learning Environments) to develop software tools for supporting middle school and high school students in analyzing data and constructing explanations with biological data. Reiser is a co-principal investigator in the NSF Center for Curriculum Materials in Science. He served as a member of the NRC panel authoring the report Taking Science to School.

Mary L. Starr is a Research Specialist in Science Education in the School of Education at the University of Michigan. She collaborates with teachers and students in elementary and middle school science classrooms around the United States who are implementing *Project-Based Inquiry Science*. Before joining the PBIS team, Dr. Starr created professional learning experiences in science, math, and technology, designed to assist teachers in successfully changing their classroom practices to promote student learning from coherent inquiry experiences. She has developed instructional materials in several STEM areas, including nanoscale science education, has presented at national and regional teacher education and educational research meetings, and has served in a leadership role in the Michigan Science Education Leadership Association. Dr. Starr has authored articles and book chapters, and has worked to improve elementary science teacher preparation through teaching science courses for pre-service teachers and acting as a consultant in elementary science teacher preparation. As part of the PBIS team, Dr. Starr has played a lead role in making units cohere as a curriculum, in developing the framework for PBIS Teacher's Planning Guides, and in developing teacher professional development experiences and materials.

Acknowledgements

Three research teams contributed to the development of *Project-Based Inquiry Science* (PBIS): a team at the Georgia Institute of Technology headed by Janet L. Kolodner, a team at Northwestern University headed by Daniel Edelson and Brian Reiser, and a team at the University of Michigan headed by Joseph Krajcik and Ron Marx. Each of the PBIS units was originally developed by one of these teams and then later revised and edited to be a part of the full three-year middle-school curriculum that became PBIS.

PBIS has its roots in two educational approaches, Project-Based Science and Learning by Design™. Project-Based Science suggests that students should learn science through engaging in the same kinds of inquiry practices scientists use, in the context of scientific problems relevant to their lives and using tools authentic to science. Project-Based Science was originally conceived in the hi-ce Center at the University of Michigan, with funding from the National Science Foundation. Learning by Design™ derives from Problem-Based Learning and suggests sequencing, social practices, and reflective activities for promoting learning. It engages students in design practices, including the use of iteration and deliberate reflection. LBD was conceived at the Georgia Institute of Technology, with funding from the National Science Foundation, DARPA, and the McDonnell Foundation.

The development of the integrated PBIS curriculum was supported by the National Science Foundation under grants nos. 0137807, 0527341, and 0639978. Any opinions, findings and conclusions, or recommendations expressed in this material are those of the authors and do not necessarily reflect the views of the National Science Foundation.

PBIS Team

Principal Investigator
Janet L. Kolodner

Co-Principal Investigators
Daniel C. Edelson
Joseph S. Krajcik
Brian J. Reiser

NSF Program Officer
Gerhard Salinger

Curriculum Developers
Michael T. Ryan
Mary L. Starr

Teacher's Edition Developers
Rebecca M. Schneider
Mary L. Starr

Literacy Specialist
LeeAnn M. Sutherland

NSF Program Reviewer
Arthur Eisenkraft

Project Coordinator
Juliana Lancaster

External Evaluators
The Learning Partnership
Steven M. McGee
Jennifer Witers

The Georgia Institute of Technology Team

Project Director:
Janet L. Kolodner

Development of PBIS units at the Georgia Institute of Technology was conducted in conjunction with the Learning by Design™ Research group (LBD), Janet L. Kolodner, PI.

Lead Developers, Physical Science:
David Crismond
Michael T. Ryan

Lead Developer, Earth Science:
Paul J. Camp

Assessment and Evaluation:
Barbara Fasse
Jackie Gray
Daniel Hickey
Jennifer Holbrook
Laura Vandewiele

Project Pioneers:
JoAnne Collins
David Crismond
Joanna Fox
Alice Gertzman
Mark Guzdial
Cindy Hmelo-Silver
Douglas Holton
Roland Hubscher
N. Hari Narayanan
Wendy Newstetter
Valery Petrushin
Kathy Politis
Sadhana Puntambekar
David Rector
Janice Young

The Northwestern University Team

Project Directors:
Daniel Edelson
Brian Reiser

Lead Developer, Biology:
David Kanter

Lead Developers, Earth Science:
Jennifer Mundt Leimberer
Darlene Slusher

Development of PBIS units at Northwestern was conducted in conjunction with:

The Center for Learning Technologies in Urban Schools (LeTUS) at Northwestern, and the Chicago Public Schools
Clifton Burgess, PI
for Chicago Public Schools;
Louis Gomez, PI.

The BioQ Collaborative
David Kanter, PI.

The Biology Guided Inquiry Learning Environments (BGuILE) Project
Brian Reiser, PI.

The Geographic Data in Education (GEODE) Initiative
Daniel Edelson, Director

The Center for Curriculum Materials in Science at Northwestern
Daniel Edelson,
Brian Reiser,
Bruce Sherin, PIs.

The University of Michigan Team

Project Directors:
Joseph Krajcik
Ron Marx

Literacy Specialist:
LeeAnn M. Sutherland

Project Coordinator:
Mary L. Starr

Development of PBIS units at the University of Michigan was conducted in conjunction with:

The Center for Learning Technologies in Urban Schools (LeTUS)
Phyllis Blumenfeld,
Barry Fishman,
Joseph Krajcik,
Ron Marx,
Elliot Soloway, PIs.

The Detroit Public Schools
Juanita Clay-Chambers
Deborah Peek-Brown

The Center for Highly Interactive Computing in Education (hi-ce)
Phyllis Blumenfeld,
Barry Fishman,
Joseph Krajcik,
Ron Marx,
Elizabeth Moje,
Elliot Soloway,
LeeAnn Sutherland, PIs.

Field-Test Teachers

National Field Test
Tamica Andrew
Leslie Baker
Jeanne Bayer
Gretchen Bryant
Boris Consuegra
Daun D'Aversa
Candi DiMauro
Kristie L. Divinski
Donna M. Dowd
Jason Fiorito
Lara Fish
Christine Gleason
Christine Hallerman
Terri L. Hart-Parker
Jennifer Hunn
Rhonda K. Hunter
Jessica Jones
Dawn Kuppersmith
Anthony F. Lawrence
Ann Novak
Rise Orsini
Tracy E. Parham
Cheryl Sgro-Ellis
Debra Tenenbaum
Sarah B. Topper
Becky Watts
Debra A. Williams
Ingrid M. Woolfolk
Ping-Jade Yang

New York City Field Test
Several sequences of PBIS units have been field- tested in New York City under the leadership of Whitney Lukens, Staff Developer for Region 9, and Greg Borman, Science Instructional Specialist, New York City Department of Education

6th Grade
Norman Agard
Tazinmudin Ali
Heather
   Guthartz Aniba
Asher Arzonane
Asli Aydin
Shareese Blakely
John J. Blaylock
Joshua Blum

Tsedey Bogale
Filomena Borrero
Zachary Brachio
Thelma Brown
Alicia Browne-Jones
Scott Bullis
Maximo Cabral
Lionel Callender
Matthew Carpenter
Ana Maria Castro
Diane Castro
Anne Chan
Ligia Chiorean
Boris Consuegra
Careen Halton Cooper
Cinnamon Czarnecki
Kristin Decker
Nancy Dejean
Gina DiCicco
Donna Dowd
Lizanne Espina
Joan Ferrato
Matt Finnerty
Jacqueline Flicker
Helen Fludd
Leigh Summers Frey
Helene Friedman-Hager
Diana Gering
Matthew Giles
Lucy Gill
Steven Gladden
Greg Grambo
Carrie Grodin-Vehling
Stephan Joanides
Kathryn Kadei
Paraskevi Karangunis
Cynthia Kerns
Martine Lalanne
Erin Lalor
Jennifer Lerman
Sara Lugert
Whitney Lukens
Dana Martorella
Christine Mazurek
Janine McGeown
Chevelle McKeever
Kevin Meyer
Jennifer Miller
Nicholas Miller
Diana Neligan
Caitlin Van Ness
Marlyn Orque
Eloisa Gelo Ortiz
Gina Papadopoulos
Tim Perez
Albertha Petrochilos

Christopher Poli
Kristina Rodriguez
Nadiesta Sanchez
Annette Schavez
Hilary Sedgwitch
Elissa Seto
Laura Shectman
Audrey Shmuel
Katherine Silva
Ragini Singhal
C. Nicole Smith
Gitangali Sohit
Justin Stein
Thomas Tapia
Eilish Walsh-Lennon
Lisa Wong
Brian Yanek
Cesar Yarleque
David Zaretsky
Colleen Zarinsky

7th Grade
Mayra Amaro
Emmanuel Anastasiou
Cheryl Barnhill
Bryce Cahn
Ligia Chiorean
Ben Colella
Boris Consuegra
Careen Halton Cooper
Elizabeth Derse
Urmilla Dhanraj
Gina DiCicco
Lydia Doubleday
Lizanne Espina
Matt Finnerty
Steven Gladden
Stephanie Goldberg
Nicholas Graham
Robert Hunter
Charlene Joseph
Ketlynne Joseph
Kimberly Kavazanjian
Christine Kennedy
Bakwah Kotung
Lisa Kraker
Anthony Lett
Herb Lippe
Jennifer Lopez
Jill Mastromarino
Kerry McKie
Christie Morgado
Patrick O'Connor
Agnes Ochiagha
Tim Perez
Nadia Piltser

Chris Poli
Carmelo Ruiz
Kim Sanders
Leslie Schiavone
Ileana Solla
Jacqueline Taylor
Purvi Vora
Ester Wiltz
Carla Yuille
Marcy Sexauer Zacchea
Lidan Zhou

8th Grade
Emmanuel Anastasio
Jennifer Applebaum
Marsha Armstrong
Jenine Barunas
Vito Cipolla
Kathy Critharis
Patrecia Davis
Alison Earle
Lizanne Espina
Matt Finnerty
Ursula Fokine
Kirsis Genao
Steven Gladden
Stephanie Goldberg
Peter Gooding
Matthew Herschfeld
Mike Horowitz
Charlene Jenkins
Ruben Jimenez
Ketlynne Joseph
Kimberly Kavazanjian
Lisa Kraker
Dora Kravitz
Anthony Lett
Emilie Lubis
George McCarthy
David Mckinney
Michael McMahon
Paul Melhado
Jen Miller
Christie Morgado
Ms. Oporto
Maria Jenny Pineda
Anastasia Plaunova
Carmelo Ruiz
Riza Sanchez
Kim Sanders
Maureen Stefanides
Dave Thompson
Matthew Ulmann
Maria Verosa
Tony Yaskulski

Genetics

Genetics was developed by the PBIS development team based, in part, on an earlier unit, *How Can I Create a More Successful Crop?*, developed at University of Illinois.

Genetics

PBIS Development Team:
Francesca Casella

Mary L. Starr

Janet L. Kolodner

How Can I Create a More Successful Crop?
Lead Developers:
Barbara Hug

M. Elizabeth Gonzalez

Special Thanks
Brian Burks
Linda Burks, Ed.D.

Pilot Teachers and Contributing Developers:
Bonnie MacArthur

Michael Novak

Keetra Tipton

Rosalie Kootz

Field-test Teachers
Stephanie Goldberg

Ruben Jimenez

Emilie Lubis

David McKinney

Jen Miller

Matt Ulmann

The development of *Genetics* and of *How Can I Create a More Successful Crop?* was supported by the National Science Foundation under grants no. 0137807, 0527341, and 0639978. Any opinions, findings, and conclusions or recommendations expressed in this material are those of the authors and do not necessarily reflect the views of the National Science Foundation.

Table of Contents

Learning Set 3

Science Concepts: *Selection, selection pressure, gene expression and the environment, extinction, adaptation, predators, prey, behavioral strategies, interpretation, Charles Darwin, species, natural selection, monocultures, artificial selection, evolution, fossil evidence for evolution, fossil record, relative dating, radioactive dating, calculating frequencies, developing a theory, modeling, simulation, careful observation, reliable data, finding trends in data, planning an experiment, making recommendations, developing explanations, using evidence to support claims, criteria and constraints, collaboration, building on the work of others.*

Learning Set 4

Science Concepts: *Cells, cell division, mitosis, cancer, asexual reproduction, sexual reproduction, fertilization, meiosis, body cells, sex cells, sex chromosomes, genetic disorders, DNA, heredity, discovery of DNA, Human Genome Project, genetic engineering, careful observation, interpretation, using a microscope, using evidence to support claims, developing explanations, criteria and constraints, making recommendations, finding trends in data, modeling, simulation, collaboration, building on the work of others.*

Introducing PBIS

What Do Scientists Do?

1) Scientists...address big challenges and big questions.

You will find many different kinds of *Big Challenges* and *Questions* in PBIS Units. Some ask you to think about why something is a certain way. Some ask you to think about what causes something to change. Some challenge you to design a solution to a problem. Most of them are about things that can and do happen in the real world.

Understand the Big Challenge or Question

As you get started with each Unit, you will do activities that help you understand the *Big Question* or *Challenge* for that Unit. You will think about what you already know that might help you, and you will identify some of the new things you will need to learn.

Project Board

The *Project Board* helps you keep track of your learning. For each challenge or question, you will use a *Project Board* to keep track of what you know, what you need to learn, and what you are learning. As you learn and gather evidence, you will record that on the *Project Board*. After you have answered each small question or challenge, you will return to the *Project Board* to record how what you've learned helps you answer the *Big Question* or *Challenge*.

Learning Sets

Each Unit is composed of a group of *Learning Sets*, one for each of the smaller questions that need to be answered to address the *Big Question* or *Challenge*. In each *Learning Set*, you will investigate and read to find answers to the *Learning Set's* question. You will also have a chance to share the results of your investigations with your classmates and work together to make sense of what you are learning. As you come to understand answers to the questions on the *Project Board*, you will record those answers and the evidence you've collected. At the end of each *Learning Set*, you will apply your knowledge to the *Big Question* or *Challenge*.

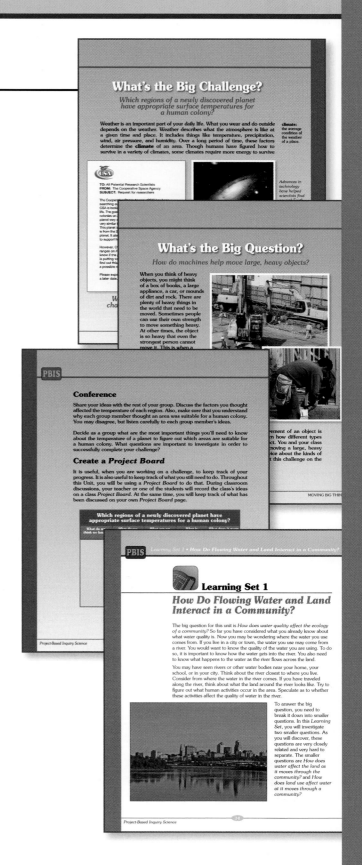

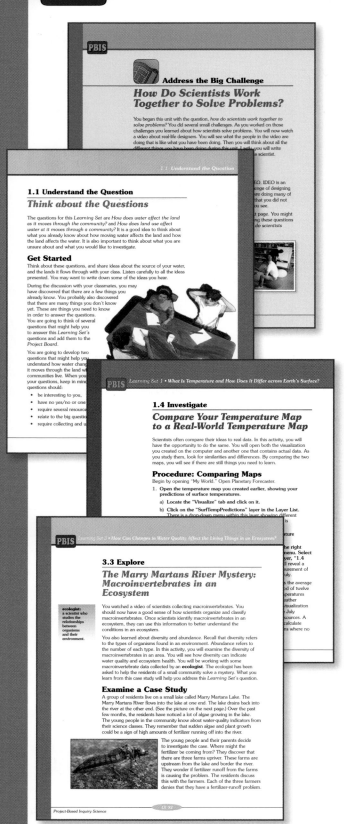

Answer the Big Question / Address the Big Challenge

At the end of each *Unit*, you will put everything you have learned together to tackle the *Big Question or Challenge*.

2) Scientists...address smaller questions and challenges.

What You Do in a Learning Set

Understanding the Question or Challenge

At the start of each *Learning Set*, you will usually do activities that will help you understand the *Learning Set's* question or challenge and recognize what you already know that can help you answer the question or achieve the challenge. Usually, you will visit the *Project Board* after these activities and record on it the even smaller questions that you need to investigate to answer a *Learning Set's* question.

Investigate/Explore

There are many different kinds of investigations you might do to find answers to questions. In the *Learning Sets,* you might

- design and run experiments;
- design and run simulations;
- design and build models;
- examine large sets of data.

Don't worry if you haven't done these things before. The text will provide you with lots of help in designing your investigations and in analyzing your data.

Read

Like scientists, you will also read about the science you are learning. You'll read a little bit before you investigate, but most of the reading you do will be to help you understand what you've experienced or seen in an investigation. Each time you read, the text will include *Stop and Think* questions after the reading. These questions will help you gauge how well you understand what you have read. Usually, the class will discuss the answers to *Stop and Think* questions before going on so that everybody has a chance to make sense of the reading.

Design and Build

When the *Big Challenge* for a Unit asks you to design something, the challenge in a *Learning Set* might also ask you to design something and make it work. Often, you will design a part of the thing you will design and build for the *Big Challenge*. When a *Learning Set* challenges you to design and build something, you will do several things:

- identify what questions you need to answer to be successful

- investigate to find answers to those questions

- use those answers to plan a good design solution

- build and test your design.

Because designs don't always work the way you want them to, you will usually do a design challenge more than once. Each time through, you will test your design. If your design doesn't work as well as you'd like, you will determine why it is not working and identify other things you need to investigate to make it work better. Then, you will learn those things and try again.

Explain and Recommend

A big part of what scientists do is explain, or try to make sense of why things happen the way they do. An explanation describes why something is the way it is or behaves the way it does. An explanation is a statement you make built from claims (what you think you know), evidence (from an investigation) that supports the claim, and science knowledge. As they learn, scientists get better at explaining. You'll see that you get better, too, as you work through the *Learning Sets*.

A recommendation is a special kind of claim—one where you advise somebody about what to do. You will make recommendations and support them with evidence, science knowledge, and explanations.

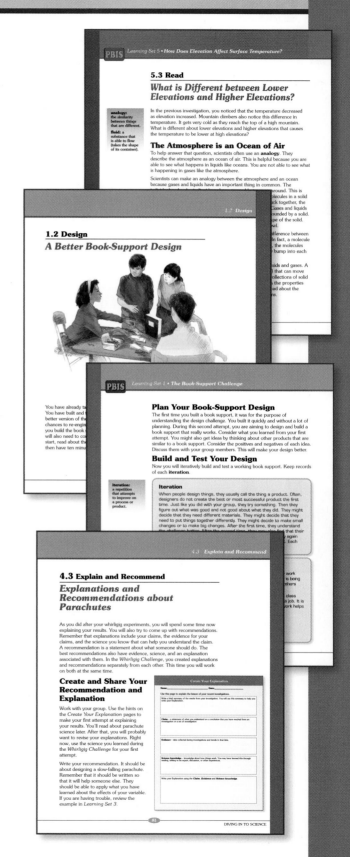

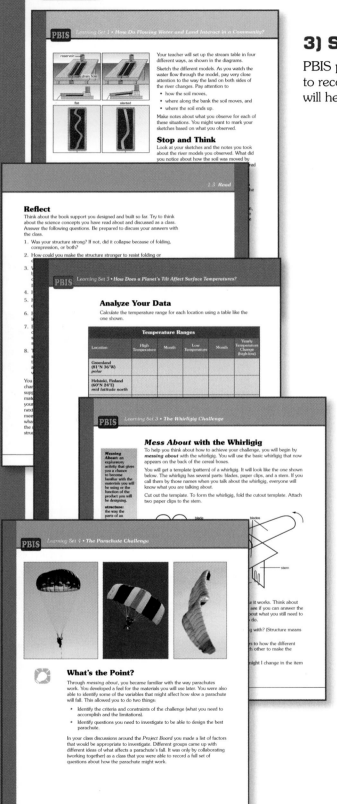

3) Scientists...reflect in many different ways.

PBIS provides guidance to help you think about what you are doing and to recognize what you are learning. Doing this often as you are working will help you be a successful student scientist.

Tools for Making Sense

Stop and Think

Stop and Think sections help you make sense of what you've been doing in the section you are working on. *Stop and Think* sections include a set of questions to help you understand what you've just read or done. Sometimes the questions will remind you of something you need to pay more attention to. Sometimes they will help you connect what you've just read to things you already know. When there is a *Stop and Think* in the text, you will work individually or with a partner to answer the questions, and then the whole class will discuss the answers.

Reflect

Reflect sections help you connect what you've just done with other things you've read or done earlier in the Unit (or in another Unit). When there is a *Reflect* in the text, you will work individually, with a partner or your small group to answer the questions. Then, the whole class will discuss the answers. You may be asked to answer *Reflect* questions for homework.

Analyze Your Data

Whenever you have to analyze data, the text will provide hints about how to do that and what to look for.

Mess About

"Messing about" is a term that comes from design. It means exploring the materials you will be using for designing or building something or examining something that works like what you will be designing. Messing about helps you discover new ideas—and it can be a lot of fun. The text will usually give you ideas about things to notice as you are messing about.

What's the Point?

At the end of each *Learning Set*, you will find a summary, called *What's the Point?*, of the important information from the *Learning Set*. These summaries can help you remember how what you did and learned is connected to the *Big Question or Challenge* you are working on.

4) Scientists...collaborate.

Scientists never do all their work alone. They work with other scientists (collaborate) and share their knowledge. PBIS helps you be a student scientist by giving you lots of opportunities for sharing your findings, ideas, and discoveries with others (the way scientists do). You will work together in small groups to investigate, design, explain, and do other things. Sometimes you will work in pairs to figure out things together. You will also have lots of opportunities to share your findings with the rest of your classmates and make sense together of what you are learning.

Investigation Expo

In an *Investigation Expo*, small groups report to the class about an investigation they've done. For each *Investigation Expo*, you will make a poster detailing what you were trying to learn from your investigation, what you did, your data, and your interpretation of your data. The text gives you hints about what to present and what to look for in other groups' presentations. *Investigation Expos* are always followed by discussions about the investigations and about how to do science well. You may also be asked to write a lab report following an investigation.

Plan Briefing/Solution Briefing/Idea Briefing

Briefings are presentations of work in progress. They give you a chance to get advice from your classmates that can help you move forward. During a *Plan Briefing*, you present your plan to the class. It might be a plan for an experiment or a plan for solving a problem or achieving a challenge. During a *Solution Briefing*, you present your solution in progress and ask the class to help you make your solution better. During an *Idea Briefing*, you present your ideas. You get the best advice from your classmates when you present evidence in support of your plan, solution, or idea. Often, you will prepare a poster to help you make your presentation. Briefings are almost always followed by discussions of your investigations and how you will move forward.

Solution Showcase

Solution Showcases usually appear near the end of a Unit. During a *Solution Showcase*, you show your classmates your finished product—either your answer to a question or your solution to a challenge. You also tell the class why you think it is a good answer or solution, what evidence and science you used to get to your solution, and what you tried along the way before getting to your answer or solution. Sometimes a *Solution Showcase* is followed by a competition. It is almost always followed by a discussion comparing and contrasting the different answers and solutions groups have come up with. You may be asked to write a report or paper following a *Solution Showcase*.

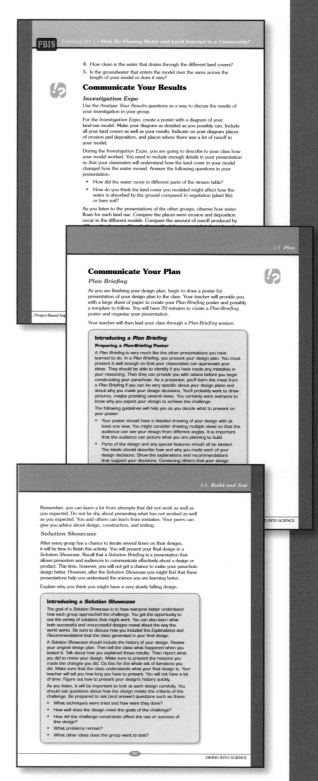

Update the Project Board

Remember that the *Project Board* is designed to help the class keep track of what they are learning and their progress towards a Unit's *Big Question* or *Challenge*. At the beginning of each Unit, the class creates a *Project Board*, and together you record what you think you know about answering the *Big Question* or addressing the *Big Challenge* and what you think you need to investigate further. Near the beginning of each *Learning Set*, the class revisits the *Project Board* and adds new questions and things they think they know. At the end of each *Learning Set*, the class again revisits the *Project Board*. This time you record what you have learned, the evidence you've collected, and recommendations you can make about answering the *Big Question* or achieving the *Big Challenge*.

Conference

A *Conference* is a short discussion between a small group of students before a more formal whole-class discussion. Students might discuss predictions and observations, they might try to explain together, they might consult on what they think they know, and so on. Usually, a *Conference* is followed by a discussion around the *Project Board*. In these small group discussions, everybody gets a chance to participate.

What's the Point?
Review what you have learned in each *Learning Set*.

Stop and Think
Answer questions that help you understand what you've done in a section.

Communicate
Share your ideas and results with your classmates.

Record
Record your data as you gather it.

GENETICS

As a student scientist, you will...

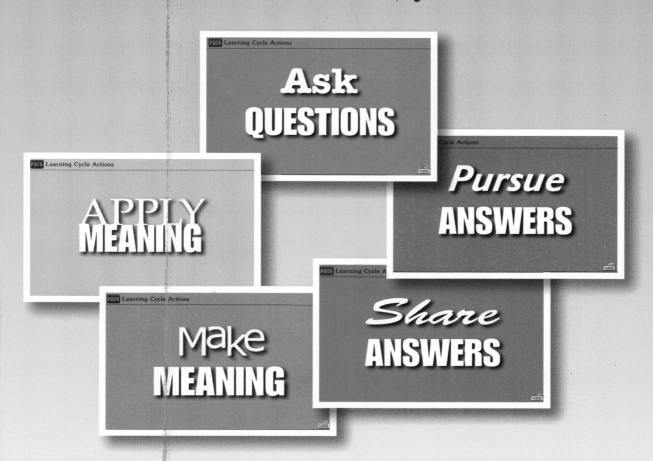

PBIS Learning Cycle Actions

Ask QUESTIONS

Pursue ANSWERS

APPLY MEANING

Make MEANING

Share ANSWERS

What's the Big Question?

How can knowledge of genetics help feed the world?

Over ten thousand years ago, humans lived as hunters and gatherers. They moved from place to place and relied entirely on wild plants they gathered and animals they could hunt for food. People then began growing plants for food. This was the beginning of **agriculture**. Learning to grow food changed the way people lived. Once people learned to plant seeds and harvest plants, they began to settle into permanent communities.

genetics: the science of how characteristics are passed down from one generation to the next.

agriculture: the production of food and other goods by growing plants and raising animals.

What Does Growing Food Have to Do with Genetics?

Wild plants changed genetically as a result of planting, harvesting, storing, and planting again. For example, some wild grass plants had seeds that clung to their stalks, while others had seeds that easily fell off during harvesting. When humans collected seeds from the wild grasses, they were able to gather more seeds from the plants with seeds that stayed on. More seeds from these plants would have been planted each year. Gradually, the plants with these genetic qualities would increase in number.

trait: a physical or behavioral characteristic of an individual that can be passed down to the next generation.

drought: a long period, lasting weeks or months, with little or no rainfall.

Over the following thousands of years, people used a similar process to transform wild plants into plants with **traits** that made the plants suitable for agriculture. Although they did not know it, these people were using genetics to grow plants for food. They did this by choosing seeds from plants that had the traits they wanted in their crops.

Most people on Earth now rely on agriculture to provide the food they need. However, farming and growing food is easier in certain places on Earth than in others. In many places, some years the environment in an area is more suitable for growing crops than in other years. Unsuitable weather conditions, such as too much rain, causing flooding, or too little rain, causing **droughts**, can make farming difficult. Insects and plant diseases can destroy crops. Sometimes people cannot grow enough food to feed themselves and their families. Farmers and scientists constantly work to find and develop plants and seeds that grow well under many different conditions.

In this Unit, you will conduct the same kinds of investigations as farmers and scientists. You will learn about genetics. You will then use genetics to make recommendations about which plants will produce the most food and the most nutritious food.

Welcome to Genetics!
Enjoy being a student scientist.

Think about the Big Question

In this Unit, you will think about how genetics might help to produce food to feed the world. You might think this problem only affects people in other parts of the world. You might not think this is something that matters to you. But everyone needs enough nutritious food to live. Growing more food and food that is more nutritious should be a concern of every person on Earth.

grain: usually a type of grass grown for its edible seeds. Grains include wheat, rice, corn, oats, barley, buckwheat, quinoa, millet, and others. Also used to describe the seed of grain plants, as in *rice grain*.

Get Started: Think about Grains You Eat

You will look at ingredient labels from several foods. Each of the foods is made from one or more common **grains**. Grains are grass plants that are grown for their edible seeds. There are many types of grains—corn, wheat, rice, rye, and barley are very common in the United Sates. Work with your group to find the grains listed on each ingredient label. Think about how important grains are to your diet.

1. Carefully read each food-ingredient label. Look for the names of the grains on each label. List each grain you find.

2. Reread each label to count the number of times each grain appears. Make a tally mark next to each grain when it appears in a list.

Hot Cereal

Nutrition Facts
Serving Size 1/4 cup dry oz. (40g)
Servings Per Container 17

Amount Per Serving Dry

Calories 140   Calories from Fat 10

% Daily Value

Total Fat 1g   2%

Saturated Fat 0g   0%

Trans Fat 0g

Cholesterol 0mg   0%

INGREDIENTS: WHOLE GRAIN WHEAT, CORN, RYE, TRITICALE, OATS, SOY, MILLET, BARLEY, BROWN RICE, OAT BRAN, FLAXSEED.

Rice Medley

Nutrition Facts
Serving Size 1/4 cup (45g)
Servings Per Container About 20

Amount Per Serving

Calories 160   Calories from Fat 15

% Daily Value

Total Fat 1.5g   2%

Saturated Fat 0g   0%

Trans Fat 0g

Cholesterol 0mg   0%

INGREDIENTS: TEX-S-MATI BROWN RICE, RED RICE, PEARLED BARLEY, RYE BERRIES.

Barley Soup

Nutrition Facts
Serving Size 1.5 cups (40g)
Servings Per Container About 3

Amount Per Serving

Calories 217   Calories from Fat 39

% Daily Value

Total Fat 5g   0%

Saturated Fat 0g   0%

Trans Fat 0g

Cholesterol 0mg   0%

INGREDIENTS: BARLEY, VEGETABLE BROTH, GARBANZO BEANS, CARROTS, CELERY, TOMATOES, GARLIC POWDER, SALT, PARSLEY, PAPRIKA, BLACK PEPPER.

Tortilla Chips

Nutrition Facts
Serving Size 1 oz. (28g/about 9 chips)
Servings Per Container 9

Amount Per Serving

Calories 140	Calories from Fat 50

	% Daily Value
Total Fat 6g	**9%**
Saturated Fat 0.5g	**3%**
Trans Fat 0g	
Cholesterol 0mg	**0%**

INGREDIENTS: STONE-GROUND ORGANIC CORN, VEGETABLE OIL (CONTAINS ONE OR MORE OF THE FOLLOWING: CANOLA, SUNFLOWER, OR SOYBEAN OIL), SALT.

Grain Bread

Nutrition Facts
Serving Size 1 slice (43g)
Servings Per Container 16

Amount Per Serving

Calories 100	Calories from Fat 15

	% Daily Value
Total Fat 1.5g	**2%**
Saturated Fat 0.5g	**0%**
Trans Fat 0g	
Cholesterol 0mg	**0%**

INGREDIENTS: WHOLE WHEAT FLOUR, STEEL CUT WHEAT, MALTED WHEAT FLAKES, CORN MEAL, ROLLED OATS, RYE FLAKES, WHEAT GLUTEN, HONEY, OAT FIBER, CONTAINS 2% OR LESS OF THE FOLLOWING: SOYBEAN OIL, BROWN SUGAR, YEAST, MOLASSES, SALT, CULTURED WHEAT FLOUR, FLAXSEED, CRUSHED WHEAT, WHEAT BRAN, SESAME SEEDS, BARLEY FLAKES, TRITICALE FLAKES, SALTED SUNFLOWER SEEDS, SOY LECITHIN, VINEGAR.

Crunchy Cereal

Nutrition Facts
Serving Size 1 cup (32g/1.1 oz.)
Servings Per Container About 9

Amount Per Serving

Calories 120	Calories from Fat 0

	% Daily Value
Total Fat 0g	**0%**
Saturated Fat 0g	**0%**
Trans Fat 0g	
Cholesterol 0mg	**0%**

INGREDIENTS: ORGANIC LONG GRAIN RICE, ORGANIC EVAPORATED CANE JUICE, ORGANIC WHOLE WHEAT, ORGANIC FREEZE-DRIED STRAWBERRIES, SEA SALT, ORGANIC BROWN RICE SYRUP, FREEZE-DRIED RASPBERRIES.

Stop and Think

1. Which grain did you find to be the most common in the ingredient labels?

2. How often do you think you eat the most common grain?

3. Which grain do you think you eat most frequently? Is it the grain that you found most common in the labels you looked at?

4. A common grain is wheat. Wheat is often used to make bread. How important do you think wheat is as a food source?

5. Another common grain is rice. Rice is eaten alone and is used in a number of other foods. How important do you think rice is as a food source?

Get Started

A Letter from the Philippines

staple food: a basic or necessary food item.

You just explored the importance of grains in your diets. Rice is an important grain and is a **staple food** for many people in the world. That means it is a basic and necessary food in their diets.

The letter on the next page is from a girl about your age. She lives in the Philippines, a country made up of about 7000 islands. The capital of the Philippines, Manila, is about 13,600 km (8600 mi) from New York City. As you read this letter, imagine the importance of rice in this family's life.

insecticide: a substance used to kill insects.

Rice harvests produce different amounts of rice each year depending on the weather, the quality of the soil, and the presence of insects and pests.

Hello Friends.

My name is Amihan. I live in the Quezon province in the Philippines. I am twelve years old. I have two brothers and two sisters. Rice is very important in our lives. We eat it three times a day. Even my favorite dessert is made with rice. Every year, on May 15, we have a festival to celebrate the rice harvest.

This year we are lucky. We have plenty of rice to eat. When there is lots of rice, my parents are happy. But last year, the harvest was not very good. My father says he cannot tell anymore when the rains will come. Sometimes they don't and then there is no rice crop.

Even when the rains come, insects may harm the rice plants. My father used to spread **insecticides**. But one day he became sick. The doctor told him that many insecticides are poisonous to people. My parents seem worried.

Two months ago, some scientists from the city came to our farm. The scientists are trying to make a new rice plant. They hope the new rice plant will grow even when the weather is bad. And insects will not eat the new rice plant, so my father will not get sick from spraying insecticides.

The scientists want my family to help them. They will give us the seeds to plant. When the new plants grow, we will tell the scientists how much rice we get.

I am happy the scientists are helping us solve the problems we are having growing enough rice.

Thank you.

Amihan

Stop and Think

The letter tells you how important rice is in the Philippines, and it explains problems people of the Philippines are having growing enough rice to keep everyone healthy. Use the letter to answer the following questions:

1. What are the problems faced by the rice farmers? List two problems rice farmers have.

2. How do you think scientists might help the farmers? Describe one way the scientists might help the farmers solve their problems.

3. How do you think farmers might help the scientists? Describe how the farmers might help the work of the scientists.

What's the *Big Challenge*?

You are just beginning to understand the importance of grains in people's diets. As you read in the letter, rice is an important grain in the diets of people in the Philippines. In fact, rice is the most important food in the diets of people around the world. More people on Earth depend on rice as their main source of nutrition than any other food.

In this Unit, you are considering a *Big Question* about providing food for people in the world. To help you answer the *Big Question*, you will also be working to achieve a *Big Challenge*. You will provide advice about developing a rice plant that is nutritious and can be grown in places that do not get a lot of rain. The letter on the next page from the *Rice for a Better World Institute* presents the challenge.

Many cultures around the world celebrate rice harvesting.

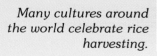

RBWI
The Rice for a Better World Institute

Research Announcement
To: All Interested Scientists
From: The Rice for a Better World Institute (RBWI)
Subject: Request for research collaboration

The RBWI is dedicated to improving rice plants to help farmers around the world grow more productive rice plants.

We have had some successes in the past, but we need to address a new and bigger problem now. The world population is growing fast. Our task is to address this situation by developing rice plants that are more nutritious and will produce more rice.

With the help of genetic technology, we have already developed several new varieties of rice. Each variety has traits that help it grow well in particular environments and under particular conditions. But for the future, we must develop new plants to address the problems rice farmers are facing. These farmers need rice plants that will grow when there is not a lot of rain.

We invite you to join the international team of researchers and farmers in the Philippines working on this project. We will keep you updated on the progress of our research from time to time. You will also be able to count on the farmers to carry out field experiments for you.

Scientists work with farmers to solve the problems of growing rice.

Identify Criteria and Constraints

criteria: goals that must be satisfied to successfully achieve a challenge.

constraints: factors that limit how you can solve a problem.

In this Unit, your challenge will be to make recommendations about developing a new rice plant that will produce more rice and more nutritious rice. Before you start, make sure you understand the **criteria** and **constraints** of your challenge. Criteria are conditions that must be satisfied to achieve the challenge. In this case, the rice plant you develop must produce more rice and more nutritious rice than the plants the farmers are now using.

Constraints are factors that limit how you can solve the problem. You cannot grow the rice in your classroom. The farmers must grow the rice in their own fields. You can only suggest ways for the scientists to work with the farmers in developing the new rice plant. The farmers can carry out any field experiments you need to conduct. Think about other constraints that may affect your solution. Record your criteria and constraints in a table like the one on the next page so you can refer to them as you move through the Unit.

Make recommendations about developing a new rice plant that will produce more rice and more nutritious rice	
Criteria	**Constraints**

Conference

Your *Big Question* for this Unit is *How can knowledge of genetics help feed the world?* Your *Big Challenge* is to advise the *Rice for a Better World Institute* about developing a rice plant that can produce more rice and rice that is more nutritious. Think about what you need to investigate to address the *Big Challenge* and answer the *Big Question*.

Working by yourself, develop two questions you have about rice. Develop two more questions you need to answer so that you can successfully answer the *Big Question* or achieve the *Big Challenge*. Remember that your questions should be interesting to you, require several resources to answer, and require collecting and using data. These should not be yes/no questions or questions with one-word answers.

When you have completed your questions, share all of your questions with the members of your group. Make sure each question meets the criteria for a good question. Reword questions that do not meet the criteria. Choose two of the most interesting questions of each kind to share with the class. Give your teacher a list with the rest of the questions so they can be used later.

Create a *Project Board*

When you are trying to answer a difficult question or solve a hard problem, it is helpful to organize your work. You will be using a *Project Board* throughout this Unit to keep track of your progress and the things you still need to do. Your class will keep a class *Project Board* and you will use your own copy of the *Project Board* for reference.

Remember that the *Project Board* has space to answer five guiding questions:

- What do we think we know?

- What do we need to investigate?

- What are we learning?

- What is our evidence?

- What does it mean for the challenge or question?

To start this *Project Board*, identify and record the *Big Question* and the *Big Challenge* for this Unit:

Big Question: *How can knowledge of genetics help feed the world?*

Big Challenge: *Make recommendations about developing a new rice plant that will produce more rice and more nutritious rice.*

What do we think we know?

In the first column of the *Project Board*, record what you think you know about the problems faced by the rice farmers and how a new rice plant might help them. How do you think rice is grown? Think about why the farmers may need a new plant to solve their problems. Why can't they continue to plant the rice they have?

Discuss what you know about the work of scientists who develop new plants. What do the scientists need to do? Don't worry if you don't know the answer. Think about what you would do if you were that scientist. Scientists work in different ways when trying to solve a problem. Discussing the different ideas they might have is important. Often, by putting together the best ideas that come up during their discussions, scientists discover a better way to address a problem.

What do we need to investigate?

Perhaps not all students in your class agree on the main problems farmers face in growing rice. Or maybe you and other members of your class have different opinions about how scientists can help the farmers. Use this column to keep track of what you would need to investigate to address the *Big Challenge*. Make sure you also record what you need to find out about rice plants and other things you are not sure about and need to find out more about.

You will return to the rest of the *Project Board* throughout the Unit. For now, work with your class to fill in the first two columns.

Learning Set 1

What Is Rice?

In this Unit, you will make recommendations about developing a new rice plant. The rice from the plant will need to be nutritious, and the plants will have to be able to produce enough rice grains even when there is too much or not enough rain. You already know that rice is a grain and that it is an important food for people in the Philippines. But you might not know a lot about rice, how it is grown, or how rice plants are different from one another. To prepare to answer the *Big Question* and address the *Big Challenge*, you will learn more about rice in this *Learning Set*. The smaller question for this *Learning Set* is *What Is Rice?* To answer this question, you will explore the parts of a rice plant, how it is grown, how one type of rice plant is different from another, and who eats rice.

Rice grows well in countries and regions with large populations and high rainfall because it requires a lot of people to cultivate it and plenty of water. Rice can be grown practically anywhere, even on a steep hill or mountain. On steep slopes, rice farmers often build a series of steps, called terraces, to make many flat surfaces for planting.

1.1 Understand the Question

Think About the Question

Begin by reading the following announcement from the *Rice for a Better World Institute.*

The Rice for a Better World Institute

Research Announcement

To: All Collaborating Scientists

From: The Rice for a Better World Institute (RBWI)

Subject: Research Update

The scientists at the Rice for a Better World Institute are eager to have your help in developing rice plants that can provide more people around the world with enough good, nutritious food.

We suggest that you begin your research by looking at what makes one type of rice plant different from another. We have stored grains of many different types of rice plants. All the grains look different, so we can see that rice grains have a variety of traits. We need to understand more about their similarities and differences. Some might be more nutritious and some might need more water than others. We would like to find out which traits make each rice plant different from the others.

Please start your investigation by learning about rice grains and rice plants. To be successful at helping us develop new rice plants, you will need to know what rice is, how it is grown, and who eats it around the world.

We are looking forward to hearing about your investigations of traits of different types of rice plants.

Conference

The *RBWI* scientists have collected many kinds of rice grains that all look different. However, the scientists do not know a lot about the traits of the rice plants. They do not know which ones are more nutritious than others or in what kinds of weather each of them grows best. With your group, discuss what you think you know about different kinds of rice plants and how rice is grown. Based on your discussion, develop four questions you need answered to learn more about how rice plants are grown and what makes rice plants alike and different from one another.

Update the *Project Board*

Your class started a *Project Board* to help you keep track of your investigations and questions about how to develop a new rice plant. To update the *Project Board*, share the questions your group developed. Record what you think you know about rice and differences among rice plants in the *What do we think we know?* column. Record your questions about rice and rice plants in the *What do we need to investigate?* column.

What's the Point?

The *RBWI* scientists want you to use genetics to develop new kinds of rice plants that will produce nutritious rice and grow when there are different amounts of rainfall. To do that, you have to understand more about rice plants and how they grow. You have generated questions about rice, and you will work on finding answers to those questions in this *Learning Set*.

1.2 Explore

What Is Rice and Who Eats Rice?

For many people in the United States, rice is not a staple food, so you might not know a lot about rice. You will need to know exactly what rice is to make your recommendations.

In the next two explorations, you will find out more about rice, how it is grown, and where in the world it is a food staple. You will work with your group on both explorations, and then the class will get back together to discuss all the things you found out and thought about.

Exploration 1: What Is Rice?

In this exploration, you will read about rice plants and how they are grown. You will make some drawings of rice plants with colored pencils based on pictures.

Materials

- **paper and pencil for drawing**
- **colored pencils**
- **paper**

Procedure

1. Begin by reading about the parts of rice plants. As you read, think about how the rice plant is similar to other plants you have read about or seen. Similarities and differences among plants will help you understand the traits of the rice plant.

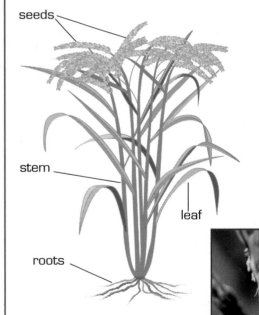

seeds

stem

leaf

roots

Rice Plants

Parts of a Rice Plant

The rice plant is a grass. It is related to wheat, oats, and barley. Like other grass plants, a rice plant has roots and stems. Leaves grow from joint-like parts along the stems. Small flowers grow on branch-like spikes at the ends of the stems. Each rice plant produces about 100 to150 tiny blossoms. These blossoms produce rice seeds.

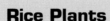

2. Before you read more, stop and sketch the roots, stems, and leaves of the rice plant. Use the picture as a guide. Make your sketch as accurate as possible. Label each part of the plant. If time permits, use pencils to color your plant to match the drawing.

3. Read about rice grains. As you read, think about the types of rice you eat.

cereal: the edible seed of a grass plant; a grain.

husk (hull): the tough outer layer on a seed.

bran: the skin of a grain.

endosperm nourishment that surrounds the germ (embryo) of a seed.

germ (embryo): the part of the seed from which a new plant grows.

Rice Grains

When you eat rice, you are eating the seeds of the rice plant. Rice seeds are a kind of grain. When a grain is edible, it is called a **cereal**. Rice grains are a kind of cereal.

Each rice grain is fairly complex, with several layers. On the outside is a tough **husk**, or **hull**. Under the hull are two more layers. First is the **bran** that protects the **endosperm**, the largest part of the rice grain. The endosperm is the food-rich part that provides the most nutrition. Finally, deep inside each grain is the **germ (embryo)**, the part of the seed from which a new plant grows.

When you eat rice, you always eat the endosperm. Other parts of the rice grain are removed in some types of rice. Brown rice still contains the bran and germ. Because of the tough outer layers, brown rice must be cooked longer than white rice. White rice is processed to remove the bran and germ, leaving only the endosperm. White rice can be stored longer and cooks more quickly than brown rice. For these reasons, most of the rice produced is processed and sold as white rice. But white rice has one important disadvantage—it is less nutritious than brown rice. Brown rice has more vitamins, minerals, and fiber.

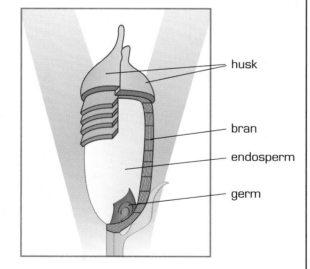

husk

bran

endosperm

germ

4. Add rice grains to your rice sketch. Make sure the grains are at the end of each stem, and label the rice grains. Use the picture of the rice plant to guide your drawing.

5. Read about how rice is grown. Then use the information you have read to answer the *Stop and Think* questions.

Rice requires a lot of water throughout its growth. When first planted, rice plants need to be submerged in water to fight weeds, so the fields must be flooded.

How Is Rice Grown?

Like other grass plants, rice has an annual cycle. Rice seeds are planted each year, generally during the spring. The seeds produce new rice plants. After about five or six months, the plants reach a height of 50–150 cm (20–60 in). During the summer, blossoms appear. Each blossom can produce a rice seed. When the rice seeds mature, they can be harvested. To produce good, nutritious seeds, most rice plants require great amounts of water, especially early in the growing cycle. Rice requires warm conditions, around 20°C (72°F) throughout the growing season. Rice also needs light for extended periods. The ideal growing conditions can be found near the Equator.

Farmers must make sure that rice plants have the correct nutrients to grow and must help protect the growing plants from pests. Farmers will often spray nutrients and pesticides on the rice plants.

Growing and harvesting rice is a difficult and complicated process involving much hard work. Machines do this work in some places, but in many places people have to do the work by hand.

Stop and Think

1. What part of the rice plant do you eat when you eat rice?

2. Could a grain of white rice grow into a rice plant? Justify your answer.

3. Planting and harvesting rice are very difficult processes. Describe how it might feel to be someone who has to plant or harvest the rice by hand. Use the photograph above to help answer this question.

Project-Based Inquiry Science

4. Using your knowledge of how rice is grown, describe what you think is the ideal environment for growing rice.

5. Using the maps of the United States and Asia, identify the places where rice is grown.

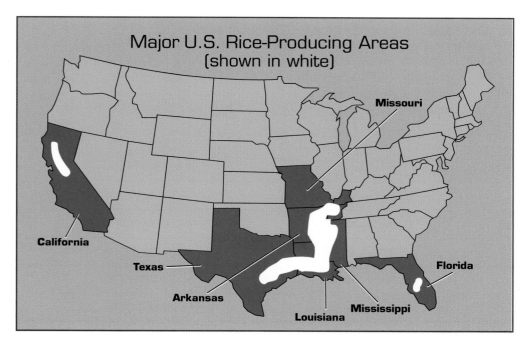

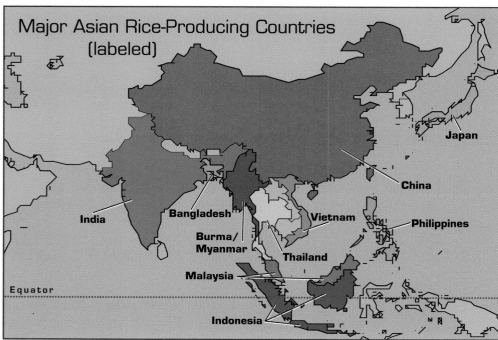

How Much Rice Do People Eat?

Rice is a staple food in most of the world, and it is the most important source of energy for much of the world's population. Much rice is consumed in the country in which it grows—for example, most of the rice grown in China is eaten in China.

This table shows the amount of rice consumed per million people in a number of countries and the amount of rice eaten per person per year in each of those countries.

Country	Total amount of rice eaten (per million people per year 2002-2003)	Amount of rice eaten per person (per year 2002-2003)
Bangladesh	183,000 metric tons	183 kg (403 lb)
Burma/Myanmar	217,000 metric tons	217 kg (478 lb)
China	103,000 metric tons	103 kg (227 lb)
India	79,000 metric tons	79 kg (174 lb)
Japan	68,000 metric tons	68 kg (150 lb)
Philippines	110,000 metric tons	110 kg (243 lb)
South Africa	15,000 metric tons	15 kg (33 lb)
United States	13,000 metric tons	13 kg (29 lb)
Vietnam	212,000 metric tons	212 kg (467 lb)

How Much Rice Do People Around the World Eat?

bar graph: a type of graph that uses either vertical (up and down) bars or horizontal (across) bars to show data. Data can be in words or numbers.

Exploration 2: Who Eats Rice?

In this exploration, you will read about how much rice is grown in different countries around the world. You will also compare the amount of rice each person eats in the different countries that grow rice. You will design a graph to display this data.

1. Look at the data table above. Think about the amount of rice eaten in the countries listed.

2. Using the data provided in the table, draw a **bar graph** showing how much rice is eaten per person in each country. Make sure you title your

graph, label each axis, and carefully draw in the bar for each country's data. You will need to decide on a scale to use for your vertical axis.

3. Use the table and your graph to answer the *Stop and Think* questions.

Stop and Think

1. Which country has the highest per person consumption of rice? Which country has the lowest?

2. Where in the world are the countries that consume the most rice located?

3. Which questions were easier to answer using the data table? Which were easier to answer using the bar graph? Give the reasons for your answers.

Update the *Project Board*

You have started to explore rice, how it grows, and who eats it. Throughout the rest of this Unit, you will use this information to make recommendations. You will also find out more about rice. Before continuing, record what you have discovered about rice, how it is grown, and who eats it, in the *What are we learning?* column of the *Project Board*. Be sure to add your evidence to the *What is our evidence?* column. Sometimes evidence can come from your reading. You can also use data from the tables and graphs as evidence.

These readings may also have brought more questions to mind. Record those questions in the *What do we need to investigate?* column so you can come back to them later.

What's the Point?

Rice grows on rice plants. When you eat rice, you are eating the seed of the rice plant. The seed of the plant is also used to grow more rice. Rice plants need a climate with lots of sunlight, water, and warm temperatures to grow well. The process of preparing the ground, planting, growing, and harvesting rice plants can be very difficult.

People around the world eat rice. It is a staple food for most people in Asia. For these people, most of their food energy comes from rice.

1.3 Explore

How Do Organisms Differ From One Another?

Rice grains are all harvested and processed in a similar way, but not all rice grains are exactly the same. To address the challenge, it will be important for you to understand differences among kinds of rice and **variations** even in the grains of rice plants of the same type.

variation: the differences among individuals in a group.

Because the idea of variation is complicated, before examining the differences among rice grains, you will start by looking at differences among an organism that is familiar to you, humans. Later, you will use what you find out about differences among humans to examine differences among rice grains.

Get Started

Many physical traits are similar across people. Other physical traits may make you different from your classmates. Working with your class, you will identify these traits and record the similarities and differences.

On a piece of paper, make two columns, one for similarities among humans and the other for differences. Discuss with your class how human beings are similar to one another in their physical characteristics.

Now, think about your physical characteristics. Try to identify traits that human beings have in common. These can be how many legs or how many eyes you have. List those in the similarities column.

Work with your class to identify traits that are different among human beings. What makes you different from your classmates? Record those in the differences column.

Conference

Working as a group, look at the photos shown on the next page. Each photo shows one human trait and two different ways that trait might appear. As you look at the traits, think about which ones you have. Can you roll your tongue? Do you have attached or detached ear lobes? Look at the photos on the page to understand how each trait varies among humans. In the My Traits row of your *Inventory of Traits* page, make a check mark in the boxes that match your traits.

Work with your group to complete the second part of the *Inventory of Traits* page. For each of the traits you investigated, record how many members of your group have each variation of that trait. Record your data in the space for My Group's Traits. Suppose two members of your group have the trait attached ear lobe. If they do, write the number 2 in the corresponding space on the page. If none of the members of your group have a specific trait, write 0 on the page.

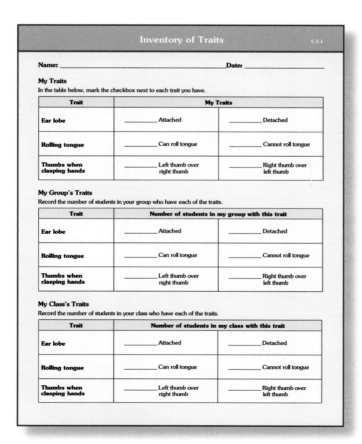

GENETICS

Human Traits

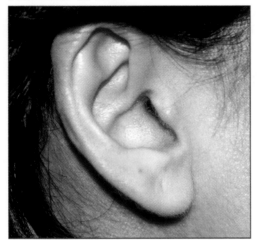

Detached ear lobe

Which variation?

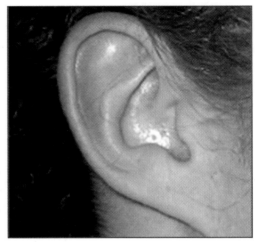

Attached ear lobe

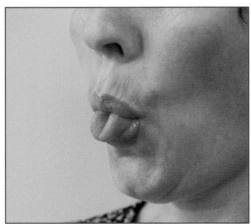

Can roll tongue

Which variation?

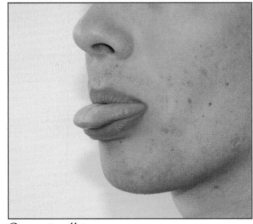

Cannot roll tongue

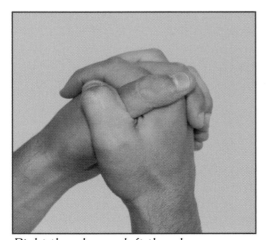

Right thumb over left thumb

Which variation?

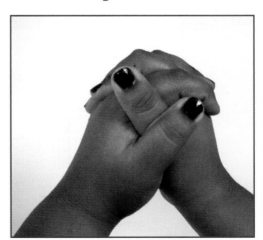

Left thumb over right thumb

Analyze Your Data

Discuss the results with your group. As you look at the data, think about the following questions:

- Which variations of each trait are shared by all your group members?

- Are there variations of the traits that none of your group members have? Which ones? Why might this be so?

- Suppose you collected data from all the students in your class instead of only from your group. How do you think your results would change with data from a larger group?

- Does any other member of your group have exactly the same combination of traits as you? Do you expect to have exactly the same combination of traits as another student? Why or why not?

- Which other traits do your group members have that might make them different from others?

- Which of the human traits you investigated do you think are common among human beings? Which ones do you think are less common?

As you discuss your answers to the questions, listen carefully as others present their ideas. Pay attention to the evidence presented by your group members to support their conclusions. Take notes during the discussion. Be prepared to present your group's data and conclusions to the rest of your class.

Communicate Your Results

Each group will have a chance to present its data to the class along with its answers to the questions. As you listen to other groups' presentations, compare your group's data with the data of other groups. How are the data similar or different? Listen carefully to other groups' answers to the questions. Groups may not agree with one another. Keep track of the variation in traits across other groups, and record data about the class on your *Inventory of Traits* page.

After all the groups have shared their data, revisit the answers to the questions. Which traits are more common or less common across the class? Is there more or less variation in traits in your class as a whole compared to your group? Using all the data you have now, work with the class to answer the questions.

Reflect

Think about rice—it also has many traits. What do rice and human traits have in common? What traits does rice have? Identify two questions about the traits of rice or rice plants. How might you investigate the answers to your questions?

Update the *Project Board*

Your class started a *Project Board* to help you keep track of your investigations and questions about developing a new rice plant. To prepare for updating the *Project Board*, share your group's questions. Based on the different human traits you have examined, update the *Project Board*. Record what you think you know about traits and differences in the *What do we think we know?* column. Record your questions about differences in types of rice and rice plants in the *What do we need to investigate?* column.

What's the Point?

Traits are individual characteristics that can be passed on to the next generation. Humans have many traits in common as well as many differences. These differences make each person unique. Like humans, rice plants show variations in traits. Scientists and farmers have asked for help in growing more rice and more nutritious rice. To develop better rice plants, scientists will need to examine many different rice traits. It will be important to understand these traits and find out how one rice plant differs from another.

1.4 Investigate

What Are Some Differences Among Rice Grains?

You have just looked at some of the differences among human beings. Now that you understand more about the variations in humans, you can begin to investigate differences among rice grains. You will do two investigations, one in which you observe rice grains to notice their similarities and differences, and one in which you measure rice grains to identify their similarities and differences.

Investigation 1: Observe Differences Among Rice Grains

What Are Some Rice-grain Traits that Can Be Observed?

Some types of rice have long grains and some have short grains. There are also other differences among types of rice. Some you can easily observe. You will be examining the grains of four different types of rice. You will examine the similarities and differences among grains of the same type of rice, and then identify the similarities and differences among grains of different types of rice.

Procedure

Work with your group to identify the traits of the rice in each container. The containers are marked A, B, C, and D, and each container has a different kind of rice grain. Use an *Observing and Comparing Rice Grains* page to record your observations.

Follow this procedure as you make your observations.

1. Fold a piece of paper into four equal parts. Label the rectangles A, B, C, and D. Put one scoop of the matching rice in each rectangle.

Materials

- **4 containers of rice labeled A, B, C, and D**

- **spoon**

- **paper and pen**

- **hand magnifying lens**

Observing and Comparing Rice Grains 1.4.1

Name: _____ Date: _____

Record your rice grain data from each cup.

	Rice description	Similarities among rice grains	Differences between rice grains	Possible reasons for similarities and differences
Rice A				
Rice B				
Rice C				
Rice D				

Compare the rice grains across different cups.

Similarities among rice grains across different piles	Differences between rice grains across different piles	Possible reasons for similarities and differences

sample: a piece or part taken from a group, whose properties are studied to gain information about the whole group.

sampling: the process of selecting a suitable sample, or representative part, of a whole group.

2. Look at the grains in each pile of rice. Within each pile, identify similarities and differences among the grains. On your *Observing and Comparing Rice Grains* page, record the similarities and differences you observe.

3. Now look at similarities and differences among rice grains in different piles. What similarities do you see? What differences do you notice? On your *Observing and Comparing Rice Grains* page, record the similarities and differences you observed.

Analyze Your Data

1. Review your observations of rice grains from containers A, B, C, and D. How different from one another are the rice grains in the A sample? How different from one another are the rice grains in samples B, C, and D?

2. Review your observations of rice grains to identify differences in the grains between containers A, B, C, and D. How large or small are the differences across the rice samples? Use evidence from your observations to support your claims.

3. What traits of rice are similar across rice types? What traits of rice are different across rice types?

Be a Scientist

Sampling

A **sample** is a piece or part taken from the whole group. When scientists investigate something, such as the size of rice seeds, they cannot possibly measure every item in the set. Instead, they measure a sample, a small number of items chosen at random. From this they can estimate the size of others. In science research, selecting a suitable sample, or representative part of the whole group, is called **sampling**.

Investigation 2: Measure Differences in Rice Grains

What Are Some Rice-grain Traits that Can Be Measured?

Looking at differences in rice grains and documenting them is one way of getting evidence. You can obtain more precise evidence of similarities and differences in rice grains by measuring the length and width of different rice-grain samples.

Procedure

Using a ruler, first measure differences in size among grains in a single rice sample. Each member of your group will analyze a different rice sample: A, B, C, or D. You will then share your data with your group and use this data to answer questions.

Materials

• **4 labeled containers of rice (A, B, C, and D)**

• **paper**

• **4 rulers**

• **transparent tape**

1. Select five grains of rice from the container assigned to you (A, B, C, or D). Select grains that are whole and not broken or chipped.

2. Tape each grain you selected to paper so it will not move while you measure it. Leave enough space between the grains so you can easily measure the length and width of each.

3. Use the ruler to measure the length and width of each of the five grains. Measure each grain to the nearest millimeter. Measure carefully by making sure you have the ruler at one end or side of the rice grain. Record these measurements on a *Variation in Rice Grains* page.

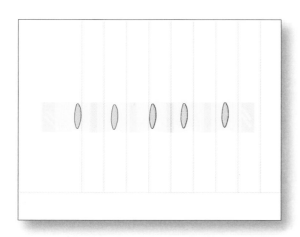

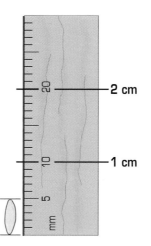

Note: Not to scale

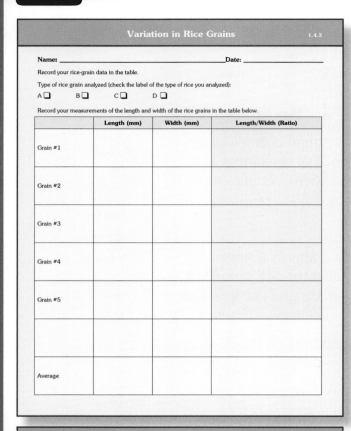

4. When you have measured five grains, calculate the average length of your rice type and calculate the average width of your rice type. To calculate averages, add up the measurements for each rice grain and divide by the number of grains you measured (5). Round your answer to one decimal place.

5. Calculate the ratio of the average length to average width for the grains of your type of rice. To calculate the ratio, divide the average length by the average width. Round your answer to one decimal place. Record this ratio in the bottom right box on the *Variation in Rice Grains Analysis* page.

Analyze Your Data

1. Begin by sharing your averages and your ratio with your group. Each group member should record all the group averages on a *Variation in Rice Grains Analysis* page.

2. Compare your data for each of the rice-grain samples. How do the average length and width of your rice sample compare to the average length and width of the other samples? Record the comparisons on your *Variation in Rice Grains Analysis* page.

3. How does the ratio of average length to average width for your sample compare to the ratios of the other samples? Graphs can help you compare numbers in a visual way. To compare the ratio of average length to average width of the different samples, you should plot a graph. Prepare a bar graph on your *Variation in Rice Grains Analysis* page. Use the *x*-axis (horizontal axis) for the sample label and the *y*-axis (vertical axis) for the ratio. After preparing your bar graph, record the comparisons on your *Variation in Rice Grains Analysis* page.

> ### Classifying Types of Rice
> Rice grains are classified based on their ratio of length to width. This classification method provides scientists and farmers with a way to compare different types of rice. For example, rice grains that are very long compared to their width would have a high length to width ratio, more than 1:1. Other rice may be almost as long as it is wide. These types of rice would be classified as having a ratio lower than 1:1.

Communicate Your Results

Meet with the class to discuss your results. Present the values your group found for the average length, average width, and length/width ratio. Compare your values with those of other groups in your class. Use the following questions to guide your discussion.

- Were there differences among the average lengths and widths for each type of rice over the different groups? If there were differences, what might be the causes?

- Were there differences among the length/width ratios for each type of rice over the different groups? If there were differences, what might be the causes?

- If needed, discuss the measurement procedures you used. As a class, decide on the most accurate values for the average length, average width, and length/width ratio of each type of rice.

- Now that you have the most accurate values from each sample of rice, are the differences among the different types of rice larger or about the same as the differences within one type of rice?

What's the Point?

Rice grains have many traits. You can observe the traits of size and color, among others. When comparing rice grains from the same type of rice plant, the traits of the grains are very much the same. Length and width measurements do not vary much across rice grains of the same type. The differences across different types of rice are larger than the differences within one type of rice. Traits that can be observed and measured are used to classify different types of rice.

1.5 Read

How Does Rice Provide Nutrition?

starch: a tasteless, odorless carbohydrate found in foods.

carbohydrate: a complex sugar. Carbohydrates provide energy when digested.

You know from your investigations in the last section that different varieties of rice grains have different traits. The size and color of the grain are two differences. Another trait that can vary is how nutritious the rice is. People eat rice because it can be very nutritious. It provides energy in the form of **starch**.

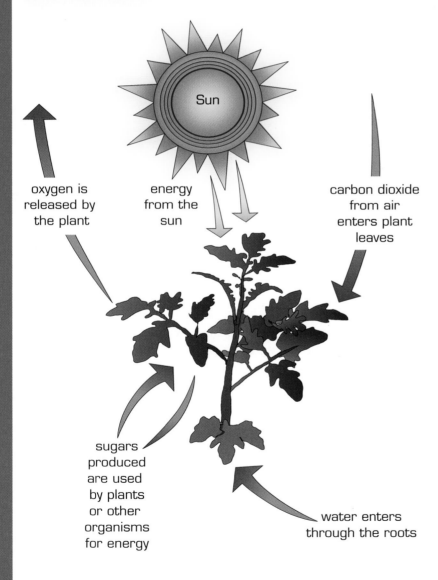

oxygen is released by the plant

energy from the sun

carbon dioxide from air enters plant leaves

sugars produced are used by plants or other organisms for energy

water enters through the roots

Sun

Starch is a type of complex sugar produced by plants. Complex sugars, are known as **carbohydrates**. When you eat carbohydrates, they provide energy for your body. Plants use energy from the Sun, along with water and carbon dioxide from the atmosphere to produce sugars in a process called **photosynthesis**. Some of the sugars plants make are used for growth and to support their life systems. The remaining sugars are stored as starch in the seeds or roots of the plant.

When humans and other animals eat plants, they consume the starch the plants produced. After the starch is eaten, the carbohydrates are digested, or broken down, in a form the animal's body can use as energy. Starch from grains is a very important source of nutrition for people all over the world. Rice provides humans with starch, minerals, vitamins, and some proteins. All of these compounds are essential for people. Because so many people in the world depend on rice for nutrition, rice is one of the most important grains among the cereals.

Today, millions of people in some areas of the world suffer from **malnutrition**. Because rice is the main food for most of the world's people, scientists want to find a rice variety that will grow in many different areas and has the best possible nutritional value. If such a rice variety can be found, it will help to feed the world.

Stop and Think

1. Why is rice a staple crop for millions of people?

2. Why might it be important to differentiate rice plants by the amount of starch they contain?

3. How is the starch found in rice grains produced?

Update the *Project Board*

While working on this *Learning Set,* you have read about how rice is grown and who eats it. You have also identified several differences among different kinds of rice grains and measured some of the variations within types of rice grains and across types of rice grains. Add to the *What are we learning?* column of the *Project Board* what you have learned about similarities and differences across types of rice plants and what you have learned about rice and nutrition. Remember to add evidence to the *What is our evidence?* column for each entry to the *What are we learning?* column. Evidence can come from your reading or from your investigations.

photosynthesis: a process in which green plants use the energy from sunlight along with carbon dioxide and water to make their own food (sugar) and oxygen.

malnutrition: a condition resulting from not enough food or lack of the proper food.

What's the Point?

Rice plants use energy from the Sun to produce sugars in a process called photosynthesis. Plants use some of the sugar they produce for growth and to support their life systems. The remaining sugars are stored as starch in the seeds and roots of the plants. When people eat rice, they digest the starch in the rice seeds, and it provides them with energy. Rice provides humans with starch, minerals, vitamins, and proteins. Rice grains contain different amounts of starch. The more starch in a grain, the greater its nutritional value. If scientists can find a rice variety that will grow in many different areas and that has high nutritional value, it will help to feed the world. Because so many people in the world depend on rice for nutrition, rice is one of the most important grains among the cereals.

Learning Set 1

Back to the Big Challenge

Make recommendations about developing a new rice plant that will produce more rice and more nutritious rice.

The Rice for a Better World Institute

To: All Collaborating Scientists

From: The Rice for a Better World Institute (RBWI)

Subject: Rice Traits

The Rice for a Better World Institute (RBWI) has received your information on how rice grains vary in size. We think you will need to know the traits of the different rice our researchers have developed. We have listed those traits in the chart below.

Rice variety	Trait
A	grows well in dry conditions
B	grows well even in flood conditions
C	has high starch content
D	has high fiber content
E	has high levels of vitamins and minerals
F	is resistant to pests
G	is resistant to disease

Rice variety	Trait
H	produces more rice grains per plant than other rice plants
I	requires less fertilizer per acre of rice than other rice plants

The goal of the RBWI and the collaborating scientists is to combine as many of these traits as possible in a new rice plant. As you go forward in your investigation, please keep this goal in mind. With your help, we may be able to produce a new rice plant with desirable traits.

Conference

Using the information the *RBWI* scientists have just sent and what you now know about rice and the traits of rice plants, discuss the answers to the following questions with your group, and identify how each relates to the criteria and constraints of the challenge.

- **What kind of traits do you think are most desirable for a new rice plant to have? How will this affect your recommendations?**

- **How will you make sure the traits of the new rice plant will meet the criteria?**

- **How will constraints you identified earlier affect your recommendations?**

- **What information from this *Learning Set* will you use to help you make your recommendations?**

- **What questions do you still have?**

- **Examine the rice traits in the letter from the *RBWI*. Identify why each might be important to growing more rice or rice that is more nutritious.**

Communicate

Share your group's answers to the questions with the class. As you listen to your classmates, make sure you understand the answers to these questions. If you do not understand something, or if they did not present something clearly enough, ask questions. Ask your questions and make your suggestions respectfully.

Update Criteria and Constraints

Revisit the criteria and constraints for this challenge. Now that you know more about how rice plants are different from one another, you may have found that there is more to think about than you earlier imagined. You may now realize that the criteria and constraints are different from what you first expected. For example, you know that the amount of starch in a seed is important in developing a more nutritious rice. You read information from the *RBWI* and now know that developing rice plants resistant to pests and diseases is also important to think about. Using your new knowledge, update your list of criteria and constraints, making it more accurate. A more accurate list will help you better achieve the challenge.

Update the *Project Board*

What you now know about different traits in rice plants has probably given you a better idea of what you need to do to address the challenge. Your new knowledge has allowed you to identify additional questions you need to answer. You might also have ideas about investigations you would like to conduct. You have updated the criteria and constraints for the challenge. Now add your new questions and ideas for investigations to the *Project Board*. Add your questions and ideas to the *What do we need to investigate?* column.

Put your recommendations about traits you think the rice plants should have in the last column of the *Project Board*. Feel free to add to the *What are we learning?* or *What is our evidence?* or *What do we think we know?* columns if you discover things that you did not put into those columns earlier. As the class *Project Board* is updated, remember to update your personal *Project Board*.

Learning Set 2

How Are Traits Passed Down From Generation to Generation?

Understanding how traits are passed down from generation to generation is important for scientists trying to develop a new rice plant. They need to make sure that any new rice plant has the traits they want it to have.

The traits of living organisms vary. Rice traits include size and amount of starch. Begin this *Learning Set* by reading the letter from the *RBWI* on the next page. The scientists have given you more details about the traits that the new rice plant should have.

Flowers from the same kind of plant can show surprising variety.

The Rice for a Better World Institute

Research Announcement

To: All Collaborating Scientists

From: The Rice for a Better World Institute (RBWI)

Subject: Research Update

Here at the Rice for a Better World Institute, we are very pleased with the progress of the research so far. We have completed the inventory of the traits of all the rice samples. We found many types of rice with different traits. We think some of these traits are better than others for our purposes. We found several types of rice that have good nutritional value. We would like to have this trait in the new plant. We also found other rice that grows well when the weather is poor. In addition, we would like to have this trait in the new plant. But the traits appear in two different plants and we need to find a way to combine them into a single plant.

Our next step is to understand how traits are passed on from one generation to the next. We therefore suggest that, for the next phase of the project, you investigate how organisms pass on traits during reproduction.

We are counting on your help to solve this problem. If you need to run field experiments, the farmers collaborating on this project will be happy to assist you.

We will keep you posted on the progress of our research. In the meantime, good luck with your investigation!

2.1 Understand the Question

Think About the Question

To help you think about how traits are passed down from one generation to the next, you will build a Reeze-ot plant. The Reeze-ot is a model, not a real plant. In many ways, though, a Reeze-ot looks like rice. It has a stem and leaves, and it produces a spike with flowers. Seeds develop from the flowers.

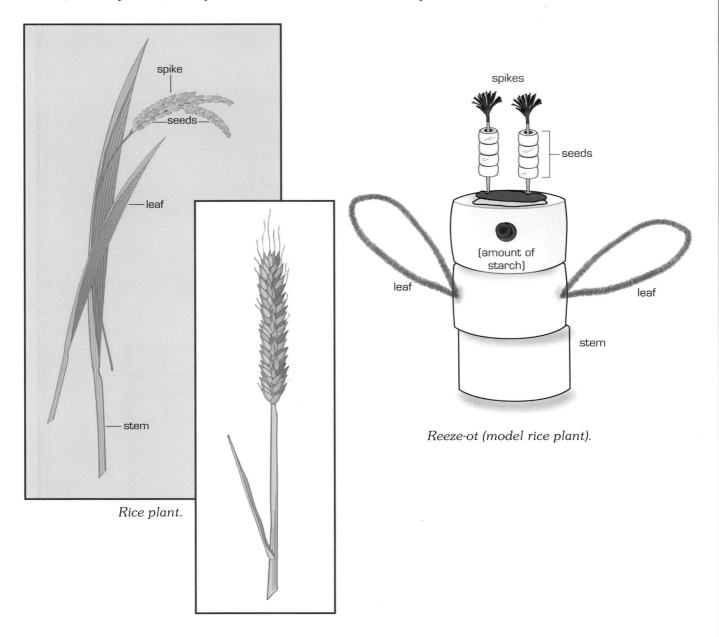

Rice plant.

Reeze-ot (model rice plant).

Get Started

You will work in pairs to build a Reeze-ot. Each Reeze-ot has eight traits. Traits are represented by pairs of letters, either two uppercase letters, two lowercase letters, or one of each. For example, the letter *H* represents the trait *height*. A Reeze-ot can have the letter combinations *HH*, *Hh*, or *hh* for the trait *height*. Letter combinations *HH* and *Hh* mean the Reeze-ot will be tall. The letter combination *hh* means the Reeze-ot will be short. You will select trait cards for each Reeze-ot trait. These cards will determine the traits your Reeze-ot will have and how those traits look on your Reeze-ot. Using the traits on your cards, you will build a Reeze-ot.

Materials

- **8 pairs of Reeze-ot trait cards**
- ***My Reeze-ot Traits* page**
- ***Key to Reeze-ot Traits* page**
- **rubber stopper**
- **marshmallows**
- **long white or green pipe cleaners**
- **party toothpicks**
- **piece of yellow clay**
- **piece of green clay**
- **white pushpin, red pushpin, or transparent pushpin**
- **white beads**

Reeze-ot Traits

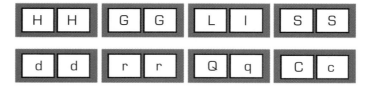

Procedure

1. The trait cards are in eight different containers. Pick one card from each container. Your partner will also select a card from each container. Combine your cards with your partner's. Match the letters on them and arrange them in pairs on the table with the letters facing up. You should have 8 pairs of cards; you selected one card from each pair, and your partner selected the other.

2. On your *My Reeze-ot Traits* page, list the letter combinations of each pair of trait cards in the box for that trait. Make sure to record the letter combinations exactly as they appear on your trait cards.

3. Using the *Key to Reeze-ot Traits*, look up each letter combination on your *My Reeze-ot Traits* page. Record on your *My Reeze-ot Traits* page how each trait will look in your Reeze-ot. Be sure to record the correct form of the trait for each letter combination.

4. You are now ready to build your Reeze-ot. Start by placing two toothpicks in the narrow end of the rubber stopper. The toothpicks will hold the marshmallows, and the rubber stopper will be a stand for your Reeze-ot.

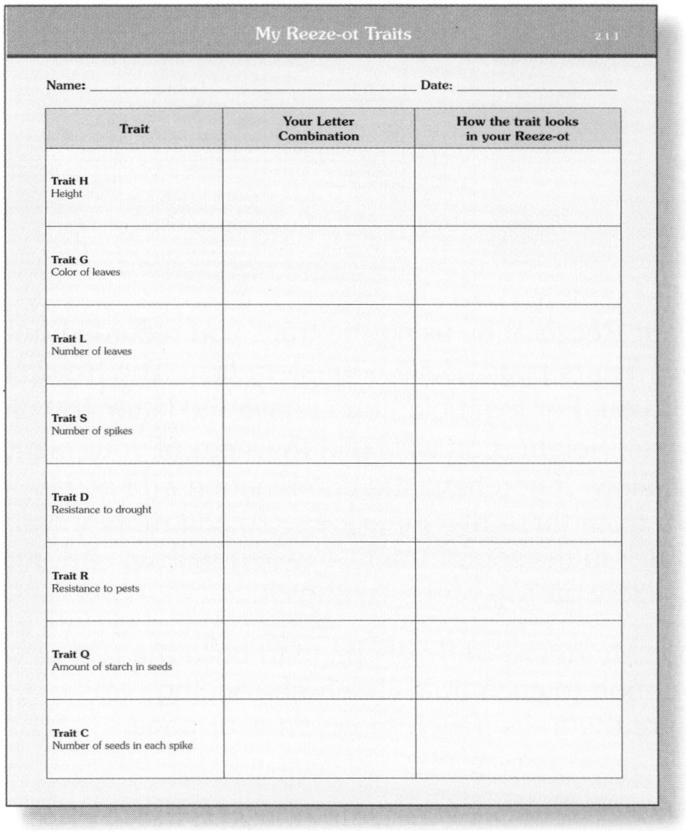

My Reeze-ot Traits 2.1.1

Name: _____ Date: _____

Trait	Your Letter Combination	How the trait looks in your Reeze-ot
Trait H Height		
Trait G Color of leaves		
Trait L Number of leaves		
Trait S Number of spikes		
Trait D Resistance to drought		
Trait R Resistance to pests		
Trait Q Amount of starch in seeds		
Trait C Number of seeds in each spike		

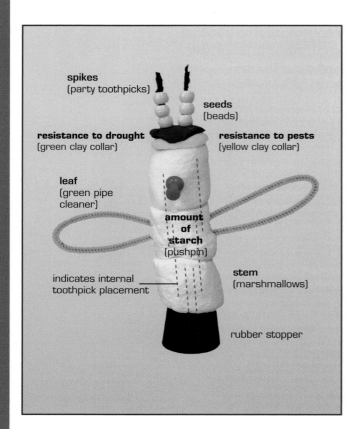

spikes
(party toothpicks)

seeds
(beads)

resistance to drought
(green clay collar)

resistance to pests
(yellow clay collar)

leaf
(green pipe cleaner)

amount
of
starch
(pushpin)

indicates internal
toothpick placement

stem
(marshmallows)

rubber stopper

Example of a complete Reeze-ot. Yours will look different.

Key to Reeze-ot Traits

Name: _____ Date: _____

Key to Reeze-ot traits

Trait	Letter Combinations		
Trait H	**HH**	**Hh**	**hh**
Height	Plant is tall. Use 3 marshmallows to build the plant's stem.	Plant is tall. Use 3 marshmallows to build the plant's stem.	Plant is short. Use 1 marshmallow to build the plant's stem.
Trait G	**GG**	**Gg**	**gg**
Color of leaves	Leaves are green. Select the green pipe cleaners.	Leaves are green. Select the green pipe cleaners.	Leaves are white. Select the white pipe cleaners.
Trait L	**LL**	**Ll**	**ll**
Number of leaves	Plant has 2 leaves. Model the leaves using 2 of the pipe cleaners. Insert the leaves into the middle of the stem.	Plant has 2 leaves. Model the leaves using 2 of the pipe cleaners. Insert the leaves into the middle of the stem.	Plant has 1 leaf. Model the leaf using 1 pipe cleaner. Insert the leaf into the middle of the stem.
Trait S	**SS**	**Ss**	**ss**
Number of spikes	Plant has 2 spikes. Insert 2 party toothpicks on top of the stem.	Plant has 2 spikes. Insert 2 party toothpicks on top of the stem.	Plant has 1 spike. Insert 1 party toothpick on top of the stem.
Trait D	**DD**	**Dd**	**dd**
Resistance to drought	Plant is not resistant. No collar.	Plant is not resistant. No collar.	Plant is resistant. Use a piece of green clay to model a collar. Place the collar around the spike, or spikes.
Trait R	**RR**	**Rr**	**rr**
Resistance to pests	Plant is not resistant. No collar.	Plant is not resistant. No collar.	Plant is resistant. Use a piece of yellow clay to model a collar. Place the collar around the spike, or spikes.
Trait Q	**QQ**	**Qq**	**qq**
Amount of starch in seeds	High. Insert a white pushpin in the stem.	Medium. Insert a red pushpin in the stem.	Low. Insert a transparent pushpin in the stem.
Trait C	**CC**	**Cc**	**cc**
Number of seeds in each spike	Large. Thread 4 white beads on the spike. Re-insert the spike in the stem.	Medium. Thread 3 white beads on the spike. Re-insert the spike in the stem.	Small. Thread 2 white beads on the spike. Re-insert the spike in the stem.

5. Build your Reeze-ot by using the traits you recorded on your *My Reeze-ot Traits* page. Build your stem first. The Reeze-ot picture will help you. For example, if you have the letter combination *hh* for the trait *height*, you will build the stem of your plant from one marshmallow. If you have the combination *HH* or *Hh*, you will build the stem from three marshmallows. To add more marshmallows, stick a toothpick in the top of the first marshmallow, and push another marshmallow on top of the toothpick.

6. Use your *My Reeze-ot Traits* page to build the rest of your Reeze-ot. Pay attention to each trait. Be careful adding parts to your Reeze-ot. Make sure the traits match those on your page.

Stop and Think

1. What similarities and differences do you expect to see when you look at all the Reeze-ots that have been built?

2. Why do you expect to see those similarities and differences?

Discuss the answers to these questions with your partner, and be prepared to present your answers to the class.

Communicate Your Results

Investigation Expo

You will display your Reeze-ot to the class in an *Investigation Expo.* Make a small poster to display along with your Reeze-ot. Include the following information on it:

- a chart showing your Reeze-ot's letter combinations and the corresponding Reeze-ot traits;

- a short paragraph answering the *Stop and Think* questions.

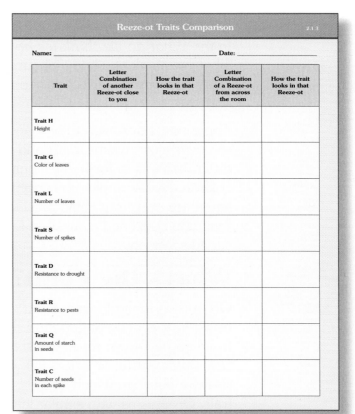

During this *Investigation Expo*, you will be walking around the room to see all the Reeze-ots that have been built. When you look at the other Reeze-ots, pay attention to the similarities and differences among them. After observing all the Reeze-ots, choose a Reeze-ot on your side of the room. Record on your *Reeze-ot Traits Comparison* page the letter combinations and traits of that Reeze-ot.

Then look at a Reeze-ot across the room from you. Record on your *Reeze-ot Traits Comparison* page the letter combinations and traits of that Reeze-ot.

As you look at the other Reeze-ots, make sure you understand which traits they have and why they have those traits.

Reeze-ot Traits Comparison

Name: _____ Date: _____

Trait	Letter Combination of another Reeze-ot close to you	How the trait looks in that Reeze-ot	Letter Combination of a Reeze-ot from across the room	How the trait looks in that Reeze-ot
Trait H Height				
Trait G Color of leaves				
Trait L Number of leaves				
Trait S Number of spikes				
Trait D Resistance to drought				
Trait R Resistance to pests				
Trait Q Amount of starch in seeds				
Trait C Number of seeds in each spike				

GENETICS

Reflect

1. When your Reeze-ot was very similar to another, how similar were the letter-trait combinations?

2. When your Reeze-ot was very different from another, how different were the letter-trait combinations?

3. Where do you think the variations in the Reeze-ots come from? How could they be so similar but so different?

Update the *Project Board*

In this section, you built a Reeze-ot with traits that matched combinations of letters. You saw similarities and differences across different Reeze-ots. You may feel that you know some things about why Reeze-ots were so similar but different from each other. You may be wondering what caused all the variation.

In the *What do we think we know?* column of the *Project Board*, record what you think you know about traits and variations. Use the *What do we need to investigate?* column to keep track of what you are not sure about and need to investigate to address the *Big Challenge*.

There may have been things you and others in the class disagreed about as you discussed similarities and differences of the Reeze-ots. Use those disagreements to form questions for the *What do we need to investigate?* column. For example, you might not have agreed about how using a pair of letters for each trait could produce so much variation. If you disagreed about that, you could add a question to the *Project Board,* asking why the Reeze-ots varied so much.

What's the Point?

Organisms look different from one another because they have different traits. In the study of genetics, traits are represented by letters, and every trait can be shown three ways — the trait can have two uppercase letters (*TT*), two lowercase letters (*tt*), or one of each (*Tt*). The traits organisms show depend on which of the combinations of letters they have for each trait. Organisms with similar letter combinations look more similar than those with different letter combinations.

2.2 Explore

How Do Flowering Plants Reproduce?

When you built the Reeze-ots, you noticed a lot of variety across different Reeze-ots. The variety, you know, came from the ways pairs of traits show themselves. However, you might not know how organisms get the traits they have.

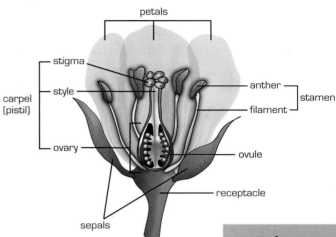

In living things, traits often come from an organism's parents. Some organisms have two parents and some (like rice and many other plants) have only one parent. When organisms **reproduce sexually**, the female parent and the male parent (or the female and male parts of the same organism) each contribute one half of the trait description (one of the letters that represent the trait). The **offspring** (child) of that reproduction shows the trait that goes with the combination contributed by the parents or parent. For example, if a mother Reeze-ot contributes an *H* for *height* and the father Reeze-ot contributes an *h* for *height*, the Reeze-ot offspring will be three marshmallows high, corresponding to the *Hh* combination.

Mammals reproduce sexually and so do flowering plants. Their flowers play a role in reproduction. Rice plants are flowering plants; they reproduce sexually. Reproduction in flowering plants includes two processes: **pollination** and **fertilization**. Pollination is delivery of **pollen**, a structure that carries the male cells, to the female part of a plant, its **ovary**. The male cells inside the pollen contain the male traits.

After pollination, fertilization takes place. A cell from inside the pollen fuses with a cell from the ovary, which contains the female traits. This fusion takes place inside the ovary, the plant structure that makes female cells.

The processes of pollination and fertilization result in seed formation. A seed receives both male and female traits. When planted, each seed can grow a new plant. The new plant will show features that are in the pairs of traits in the seed.

reproduce sexually (sexual reproduction): reproduction that combines in the offspring the female and male traits of the parents or parent.

offspring: the descendants of a person, animal, or plant.

pollination: the delivery of pollen to the female part of the plant.

fertilization: the fusion of the cell that contains the male traits with the cell that contains the female traits.

pollen: a structure in flowering plants that has cells that contain the male traits.

Materials

- newspapers or paper towels
- a flower
- hand lens
- tweezers
- scissors

ovary: the part of the plant that makes the cells containing the female traits.

carpel: the female part of flowers.

stigma: the top part of the carpel where the pollen is deposited.

style: the slender, tubelike part of the carpel.

ovule (egg cells): a tiny, egg-like structure in flowering plants that contains the female traits and develops into a seed after fertilization.

pistil: the female reproductive organ of a flower; may be made up of a single carpel or of two or more fused carpels.

stamen: the male part of the flower.

The Female Parts of a Flower

The female part of the plant is the **carpel**. The carpel includes the **stigma**, a tube-like style, and an ovary. The ovary contains **ovules (egg cells)** that contain the female traits. Fertilized ovules develop into seeds. In some flowers, the parts of the carpel are fused together. The female part of the plant is called the **pistil** in these plants.

The Male Parts of a Flower

The male part of the flower is called the **stamen**. It includes the **anthers**, where pollen (which contains **sperm cells**) is produced. Sperm cells contain the male traits. A filament, a very delicate, thread-like structure, attaches the anther to the base of the flower. Each flower has several anthers arranged around the carpel.

The Outer Covering of a Flower

Flowers have an outer covering to protect their male and female parts. The outer protective covering includes the **petals** and the **sepals**. These parts are also used to attract insects and birds that help pollinate flowers. All of a flower's parts are attached to the **receptacle** at the base of the flower.

Not all flowers look the same. Some flowers have both male and female parts. Other flowers have only the male part or only the female part. Flowers also have different numbers of sepals and petals.

Procedure: Observe the Reproductive Structures of a Flower

You will receive a flower to dissect. You will identify the male and female parts of the plant. As you do that, you should think about the role of each of these parts in reproduction and the passing on of traits.

Working with your group, spread newspapers or paper towels on the desk. Place the flower on the paper.

1. Observe the structure of the flower. Identify and count the sepals and petals. Keep in mind that the sepals and petals vary from flower to flower. The number, color, and shape of the sepals and petals in your flower may look different from the drawing showing the parts of a flower. Record the number, color, and shape of the sepals and petals in your flower.

2. Use the tweezers to detach all the petals from one side of your flower to expose its internal structure. On a piece of paper, draw the inside of your flower. Label the parts you can identify.

3. Remove a stamen. Use your magnifying glass to inspect the stamen. Can you find the pollen? Draw and describe what it looks like.

4. Use your magnifying glass to look at the end of the pistil. This is the stigma. Rub a stamen onto the stigma and, using the magnifying glass, observe it again. Record your observations.

5. Use the scissors or your fingernails to open the ovary. Use the magnifying glass to locate and count the ovules inside. Draw a diagram of the inside of the ovary.

Stop and Think

1. On your diagram, identify the parts of your plant that are used in pollination.

2. On your diagram, identify the parts of your plant that are used in fertilization.

3. Describe the process through which pollination and fertilization work to pass on traits.

anther: the structure on the stamen of flowers where pollen and sperm are produced.

sperm cells: structures that contain the male chromosomes.

petal: a flower's outer protective covering, usually colored. Used also to attract insects and animals for pollination.

sepal: a flower's outer protective covering, usually green.

receptacle: the main stem of a flower.

cross-pollination: the transfer of pollen on one plant to the female part of another plant.

self-pollination: the transfer of pollen on one plant to the female part of the same plant.

More on Pollination

You might wonder how flowers become pollinated. Some plants become pollinated when a bird, insect, or the wind transports pollen from the male part of one plant to the ovary of another plant. This is called **cross-pollination**. In other types of plants, the male part of the plant pollinates the female part of the same plant. In these plants, the male and female parts are close enough to each other so that pollen is easily transferred from the male part of the plant to the ovary of the same plant when the wind blows gently. This is called **self-pollination**.

Rice uses self-pollination for reproduction. Pollen is transferred from the stamen to the pistil of one flower. The seeds and plants produced by self-pollinating plants inherit their traits from only one parent. Therefore, new plants produced through self-pollination are more similar to the parent plant than plants produced through cross-pollination are to their parent plants.

One disadvantage of self-pollination is that the new plants are very similar to one another. If new diseases or pests are introduced or enviromental changes occur, plants with less variety of traits have less chance of survival than plants produced through cross-pollination, which have a greater variety of traits. In rice, cross-pollination is difficult, because the pollen grain lives for only a few minutes.

Bees help transfer pollen in plants that cross-pollinate. They deliver the pollen, which contains the first flower's male traits, to the second flower's female part.

Update the *Project Board*

You can now add your new knowledge from your explorations of plant reproduction to the *Project Board*. Record information about plant reproduction in the *What are we learning?* column of the *Project Board*. As you do this, remember that you must support your science knowledge with evidence. Put evidence from your reading and investigation in the *What is our evidence?* column. You may have new questions about reproduction and how plants grow from seeds. Add any new questions you might have or ideas for investigations to the *What do we need to investigate?* column. Add to the *What do we think we know?* column all the information you think you now know about how traits are passed on through reproduction.

What's the Point?

You dissected a flower to learn about the reproductive parts of a flowering plant. The ovary is the structure where the egg cells containing the female traits are produced. The anthers on the stamen are the structures where the sperm cells containing the male traits are produced. Pollen, which contains sperm cells, and ovules, which are egg cells, are both necessary for reproduction in flowering plants. The sperm cell and egg cell fuse together during fertilization to produce a seed.

The mixing of traits from the male cells and traits from the female cells produces variety in traits in the next generation. Cross-pollination occurs when the pollen of one plant is transferred to the female part of another plant. Self-pollination occurs when the pollen on one plant is transferred to the female part of the same plant. Self-pollinating plants have less variation in traits than those that produce offspring through cross-fertilization. Rice is a self-pollinating plant. In rice, cross-pollination is difficult, because the pollen lives only for a few minutes.

2.3 Read

How Do Scientists Study Traits?

Gregor Mendel and his Garden Peas

In the previous section, you observed the parts of a flower. Flowers are the reproductive organs of many plants, including rice. You also read about how flowering plants reproduce through pollination and fertilization. Gregor Mendel, an Austrian monk (a member of a religious order), was also aware of these processes when, back in the nineteenth century (1800s), he started asking questions about how plants **inherit** their traits.

Mendel had studied science and mathematics at a university before he began working in the monastery and teaching high school. As part of his duties in the monastery, he was in charge of looking after the garden. That is when Mendel began asking questions about **inheritance**. He then did experiments with ordinary garden peas to answer his questions. Like rice, garden peas are flowering plants.

Mendel studied inheritance in pea plants.

inherit: receive traits from previous generation.

inheritance: the passing down of traits from one generation to the next.

Characteristics of Pea Plants

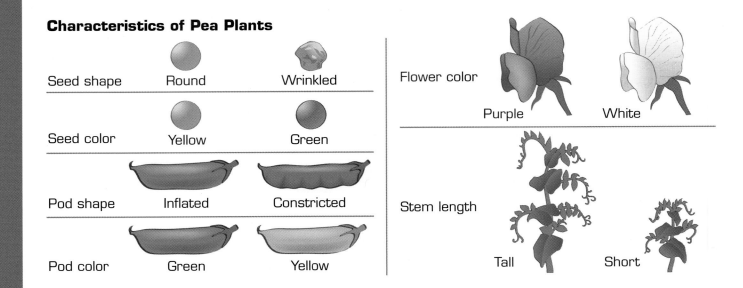

Some of the characteristics Mendel studied in his experiments.

expressed (to express): shown (to show).

blending: an equal mixing of traits.

cross: to breed two different varieties of plants to produce offspring with a mixture of traits from the two parents.

theory: a broad explanation that is strongly supported by a body of evidence.

Mendel chose to study garden peas because of the way they reproduce. He noticed that garden peas have several different characteristics. Each of these characteristics was **expressed** two different ways. Some pods from the pea plants had green seeds and some had yellow seeds. Some had round seeds and some had wrinkled seeds. None of the pods were greenish yellow, and none of the seeds were both round and wrinkled.

In Mendel's time, most people believed in the **theory** of inheritance by **blending**, an equal mixing of traits. Many of the scientists of that time believed that when you **cross**, or breed, a plant with a red flower with a plant with a white flower, the flowers of the offspring (resulting plant) would always be pink. But blending did not explain the traits Mendel observed in garden peas. Mendel did not observe any greenish-yellow pods or any seeds that were both round and wrinkled.

The other reason Mendel chose to study garden peas had to do with the way they reproduce. Rice is self-pollinating, but garden peas are both self-pollinating and cross-pollinating. Mendel could fertilize the peas by artificially cross-pollinating them. He did that by removing the anthers from one plant. He then transferred the pollen of another plant to the stamen of the plant that had the anthers removed. In this way, he could control which plants he was crossing.

Artificial Cross-pollination

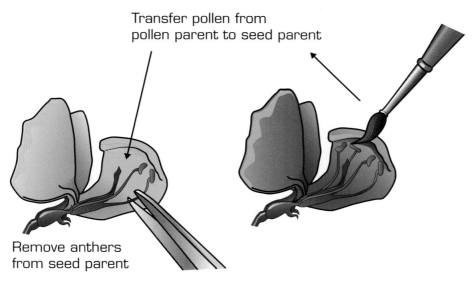

Transfer pollen from
pollen parent to seed parent

Remove anthers
from seed parent

The process of artificial cross-pollination.

Mendel's Experiments

Mendel began with the question: How are traits inherited? Because he was a scientist, he designed scientific experiments to answer his questions and made predictions as to what he thought the results would be. Then he compared his predictions to his actual results. To make his work easier, he broke his original question into smaller questions. How is plant height inherited? How is plant seed color inherited? How is plant seed shape inherited? He had many questions he wanted to answer, and he designed experiments for each of them. For each experiment, he made a prediction.

Mendel had a supply of plants that were **true-breeding**. He had observed that these plants always passed their traits on to the next generation. For example, true-breeding tall pea plants would always produce tall pea plants. In Mendel's first experiment, he asked the question: *What would happen if I bred a true-breeding tall pea plant with a true-breeding short pea plant to produce a* **hybrid** *of the two different plants?* His prediction was that all the hybrids would be medium height. That was what should have happened according to the theory of blending. But they were not. The plants were all tall. Mendel wondered what had happened to the trait for short pea plants.

true-breeding:
organisms that always pass their traits on to the next generation.

hybrid: The offspring of the cross between parents with different traits.

He thought about his results and came up with another question: *What would happen if I bred the tall hybrids with each other?* Mendel predicted that they would all be tall. But much to his surprise, some of the hybrids from this crossing were tall and some were short. The trait for shortness had reappeared.

This diagram shows the results of Mendel's experiments. When Mendel crossed true-breeding tall plants with true-breeding short plants, all the offspring were tall. When the tall plants from that cross were crossed again, most of the offspring were tall but a few were also short.

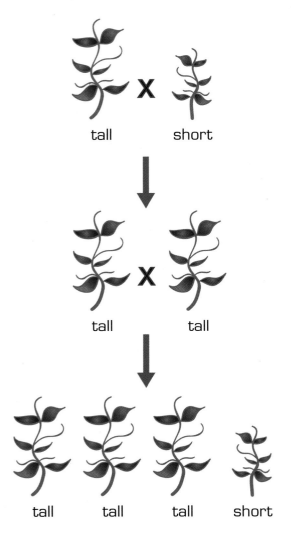

Mendel did not know what scientists know today about inheritance. Therefore, he repeated his experiments and made his observations many times. He came to the conclusion that there must be two factors for each trait, and he realized that, in some plants, both the factors were for tall plants. This would be the case for true-breeding tall plants. True-breeding short plants, he thought, would then have factors for shortness. Other plants might have one factor for tallness and one for shortness.

Mendel's experimental results showed that when one of his pea plants had two different factors for a trait, one of the factors would mask the other. For example, for plant height, he found that tallness would always mask shortness. A tall plant that had one factor for tallness and one for shortness would always be tall.

Mendel, the Father of Modern Genetics

In 1865, Mendel reported his discoveries to a scientific society, challenging the theory of blending. Unfortunately, his paper was read by very few people. People of his generation did not understand his work and were not interested in the questions he was answering. Years later, the results of his experiments in the monastery garden became the starting point of modern genetics. He is known as the "father of modern genetics."

Reflect

Answer the following questions in your group and prepare to discuss the answers in your class. Use evidence from your reading and your Reeze-ot-building activity to support your answers.

1. Mendel was a scientist who lived a long time ago. But Mendel used the same processes scientists use today. What scientific processes did Mendel use to make his discoveries?

2. Why do you think Mendel saw more variation as he crossed more plants and produced more generations of plants?

3. How would you carry out crosses to find out if a certain trait in a plant masked another trait?

4. How would you carry out crosses to find out if a certain trait in a plant was masked by another trait?

chromosomes:
strands of genetic
material inside the
cell that contain
information that
codes for the traits
of an organism.

genetic material:
genetic information
in an organism that
is passed down
from generation to
generation.

gene: the
location on the
chromosome
that contains the
instructions for a
particular trait.

alleles: different
forms of a gene.

dominant: the
allele that masks
the expression of
the recessive allele.

recessive:
the allele whose
expression is
masked by the
dominant allele.

How Do Scientists Today Explain Mendel's Discoveries?

It may surprise you, but you probably know more about genetics than
Mendel did. That is because scientists have built on Mendel's ideas and have
discovered much more about how traits are inherited than Mendel knew.

Today, scientists know that every cell in a plant or animal has **chromosomes**,
strands of **genetic material** that determine what traits the organism has.
Each chromosome contains **genes** for particular traits. People, for example,
have genes for eye color, hair color, whether they can roll their tongues, and
so on. Pea plants have genes for traits like height and shape of seeds. Each
gene is located at a particular place on a chromosome.

Position of gene for trait *Height*

One pair of chromosomes can carry different alleles for the same trait.

Allele for *tallness*

Allele for *shortness*

Scientists know that chromosomes come in pairs. Chromosomes can
have more than one form of a gene. These forms are called **alleles**. The
combination of alleles determines how a trait will express, or show, itself.
For example, Mendel observed that pea plants can be short or tall. The
alleles for pea-plant height are *T* for tallness and *t* for shortness. Because
chromosomes come in pairs, a pea plant can have two alleles for tallness
(*TT*), two alleles for shortness (*tt*), or one for each (*Tt*). One allele is on
each chromosome.

When a pea plant has two different alleles for the same trait (*Tt*), the allele
for tallness is **dominant** over the allele for shortness. Scientists call the
allele for shortness **recessive**. When a recessive allele is paired with a
dominant allele (*Tt*), the dominant allele is always expressed. When Mendel

crossed true-breeding tall plants (*TT*) with true-breeding short plants (*tt*), all the offspring were tall (*Tt*), because the allele for tallness is dominant, and the allele for shortness is recessive.

There are also other terms scientists use to describe heredity. **Phenotype** refers to the traits you can see. It is a description of what the organism looks like. When Mendel looked at the pea plants, he saw plants of different heights and peas with different shapes. You looked at traits of people and saw detached ear lobes and tongue-rolling. These are all examples of phenotypes.

Genotype refers to the actual genes and alleles an organism has. You can see a phenotype, but you cannot see the genetic makeup (the genotype) of an organism. For example, a tall pea plant could have two different genotypes. It could have two alleles for a tall stem (*TT*), or it could have one allele for a tall stem and one allele for a short stem (*Tt*). The allele for tallness is dominant, and the allele for shortness is recessive. Therefore, the plant's phenotype is tall and the genotype is either *TT* or *Tt*. A short plant can have only one genotype. Both alleles must be for a short stem (*tt*), because the tall allele (*T*) is dominant over the short allele (*t*). Though phenotypes can be seen, genotypes can be determined only through breeding experiments.

Scientists use symbols, usually letters, to represent the genotype of an organism. The symbols help them keep track of the genes and alleles of an organism. When letters are used, each gene is given a different letter. For height in pea plants, scientists use the letter *T*. Uppercase letters (*T*) represent dominant alleles, and lowercase letters (*t*) represent recessive alleles. Every gene is represented by two letters, because each chromosome of a pair has one allele for the trait. The letters can be the same, for example, *TT* for tallness or *tt* for shortness, or different letters, *Tt*, which is a tall plant. When organisms have two of the same alleles (*TT* or *tt*) for a trait, scientists say they are **homozygous**. When they have two different alleles for a trait (*Tt*), they are called **heterozygous**.

When Mendel looked at the seeds of pea plants (the peas), he noticed that many plants had round peas and fewer plants had wrinkled peas. By crossing the plants with different pea seeds, he found that round pea seeds were dominant over wrinkled pea seeds. If scientists today used "*R*" for pea-seed shape, the round pea seeds (dominant) would be either *RR* or *Rr*, and the wrinkled pea seeds (recessive) would be *rr*.

phenotype: the physical characteristics of an organism.

genotype: the genetic makeup of an organism.

homozygous: when an organism has two identical alleles for a particular gene.

heterozygous: when an organism has two different alleles for the same gene.

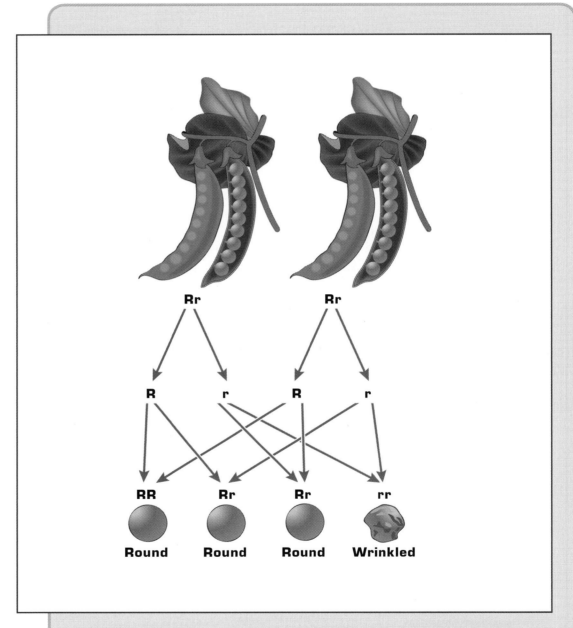

The diagram shows the cross between two heterozygous pea plants (Rr). Because the trait for round seeds is dominant, both plants have round seeds. Most of the offspring of the cross are round. The genotype for a round seed is either RR (homozygous) or Rr (heterozygous). Some of the offspring have wrinkled seeds. The genotype for wrinkled seeds can only be rr (homozygous).

Stop and Think

Answer the following questions. Be prepared to discuss your answers with the class.

1. How do a chromosome, a gene, and an allele differ? How are they similar?

2. Short stems are recessive in pea plants. If a pea plant is short, is it homozygous or heterozygous? Justify your answer.

3. Green pod color, *G*, is dominant over yellow pod color, *g*. What genotypes are possible for the phenotype with green pods?

Update the *Project Board*

In this section, you read about Gregor Mendel and his experiments about the inheritance of traits in pea plants. You also read about what scientists today know about Mendel's discoveries. Record this new science knowledge in the *What are we learning?* column of your *Project Board*. Add evidence for this new knowledge in the *What is our evidence?* column. You may think you now know more about how traits are inherited. Record what you think you know in the *What do we think we know?* column. What you read may have made you think of other questions you might have. Record these questions and ideas for further investigations in the *What do we need to investigate?* column.

What's the Point?

Gregor Mendel, in the nineteenth century, challenged the belief that traits just blended together in offspring. In his study of pea plants, he found two factors for each trait—one strong and one weak. True-breeding plants had two of the same factor—true-breeding tall plants would always produce tall plants. Plants that were not true breeding had one of each of the factors, and they would produce some tall plants and a few short plants.

Now scientists know that genes, which are locations on chromosomes, contain all the traits of an organism. The strong and weak traits are called alleles, and each gene has at least two alleles. The strong allele is called dominant. The weak allele is called recessive.

More to Learn

What Are Other Ways Traits Are Inherited?

incomplete dominance: one allele of a pair cannot completely hide the traits of its partner; this results in a "blending" of traits in the offspring.

Mendel limited his experiments to clear-cut cases of dominant or recessive traits. Not all inheritance is like this. Scientists now know of many other types of inheritance. One is very much like the old theory of blending. In **incomplete dominance,** one allele of a pair does not completely hide the other. Instead, the alleles work together to produce a different product. Snapdragon color is an example. When you cross red snapdragon flowers with white snapdragon flowers, the offspring are all pink flowers in the first generation. But unlike what the theory of blending would predict, the results of the second generation show a lot of variety, including red, white, and pink flowers.

Pink snapdragons result from incomplete dominance. The diagram shows what happens when a snapdragon with a red flower (RR) is crossed with a white flower (rr). Neither the allele for red (R) nor the allele for white (r) can completely mask the other. The result is all pink flowers. When snapdragons with pink flowers from the first generation are crossed, they can produce offspring with white (rr), red (RR), or pink (Rr) flowers .

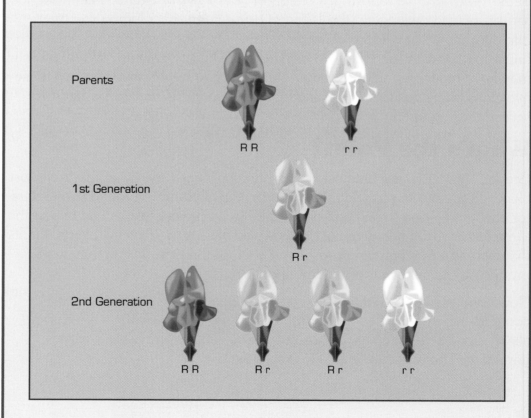

Parents   R R   r r

1st Generation   R r

2nd Generation   R R   R r   R r   r r

Another type of inheritance is **co-dominance**. In this case, neither allele of the gene masks the other. In other words, both are dominant. An example of co-dominance is in human blood types. Up until now, you have been looking at traits that are a result of only two alleles. A number of traits are the result of more than two alleles. These traits are said to have **multiple alleles**. This does not mean that an individual organism has more than two alleles for the same gene. It means that more than two alleles exist.

co-dominance: neither allele of a pair hides the other. Both alleles are dominant to the same extent; no blending occurs.

multiple alleles: more than two alleles for the same gene.

Father type B blood   Mother type A blood
(BB)   **(AA)**

Child type AB blood
(AB)

Father type A blood   Mother type O blood
(AA)   **(OO)**

Child type A blood
(AO)

A father homozygous for type A blood (AA) and a mother homozygous for type O blood (OO) will produce offspring with type A blood (AO).

A father homozygous for type B blood (BB) and a mother homozygous for type A blood (AA) will produce offspring with type AB blood (AB).

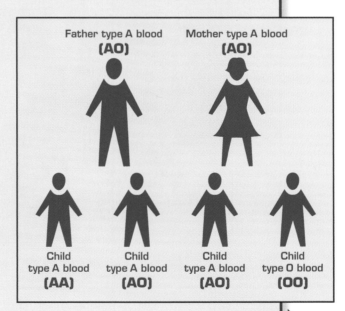

Father type A blood   Mother type A blood
(AO)   **(AO)**

Child type A blood
(AA)

Child type A blood
(AO)

Child type A blood
(AO)

Child type O blood
(OO)

A father heterozygous for type A blood (AO) and a mother heterozygous for type A blood (AO) can produce offspring with type A blood (AA) and (AO) and offspring with type O blood (OO).

The gene for blood type has three alleles: A, B, and O. The A and B alleles are dominant; the O allele is recessive. People with the genotypes AO and AA have type A blood; those with the genotypes BO and BB have type B blood; and those with the genotype AB have type AB blood. For example, a parent with type A blood and the genotype AA, and a parent with type B blood and the genotype BB, will produce children with type AB blood.

Reflect

Discuss the following questions in your class:

1. What is the difference between first and second generation offspring when characteristics are inherited through incomplete dominance? Think about the snapdragons. How does incomplete dominance make predicting the outcome of crossing plants more difficult?

2. Some science happens by chance; sometimes scientists get lucky. Mendel studied traits that were inherited only through pure dominance. How difficult would Mendel's task have been if he had chosen a trait that was inherited through co-dominance? Why?

The coat color in calico cats results from multiple alleles.

2.4 Explore

Punnett Squares and Probability

In the previous section, you read about how Mendel crossed many pea plants with different traits. After each part of his experiments, he carefully recorded the outcomes of the cross. He found that when he crossed true-breeding tall plants with true-breeding short plants, all the offspring were tall. When he bred the offspring of this cross, about ¾, or 75 percent, of the plants were tall. About ¼, or 25 percent, of the plants were short. These results led him to believe that a mathematical principle could be used to predict the outcomes of genetic crosses. He realized he could use the mathematical principle of **probability** to make predictions about the outcomes. When he bred true-breeding tall plants with true-breeding short plants, chances were that all, or 100 percent, of the plants would be tall. When he bred the resulting plants, chances were that 3 out of 4, or 75 percent, would be tall, and 1 in 4 plants, or 25 percent, would be short.

probability: the chance that something will happen.

Punnett square: a tool scientists use to investigate the possible combinations of genetic crosses.

The Punnett Square

The **Punnett square** is a tool scientists use to investigate the possible combinations of genetic crosses. It was designed by Reginald Punnett, a British geneticist. The Punnett square is a diagram that shows all the possible combinations of alleles from the female and male parents. Using the Punnett square, you can predict the possible genotypes of the offspring of a cross. From the genotypes, you can then determine the phenotypes. A simple Punnett square that looks at only one gene is a box with four squares. The top and left sides are labeled with the allele pairs of each parent (their genotypes). In the example shown, both parents are heterozygous for tallness (*Tt*). They each have the dominant allele for tallness, represented by *T*. They also have a recessive allele for shortness, represented by *t*. The table is filled in the way you would fill in a multiplication table. The alleles at the top and sides are combined to show the alleles for the offspring. One allele from each parent is passed to each offspring.

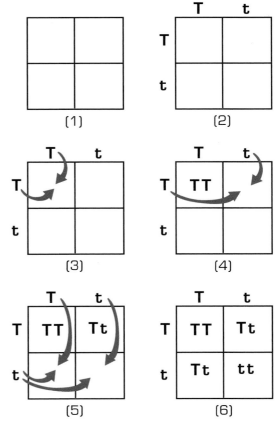

By using Punnett squares, you can calculate the probablility of the presence of a specific allele in a population of organisms. You can also see how that probablility of the allele changes as the population changes over time.

The genotypes in the boxes show the genotypes of the offspring these parents can have. You can predict the probability of offspring having each genotype by computing the fraction of each resulting genotype. The resulting genotypes of this Punnett square, showing two parents heterozygous for tallness, are

- ¼ (25%) of the offspring would probably be homozygous for tallness (*TT*)

- ½ (50%) of the offspring would probably be heterozygous for tallness (*Tt*)

- ¼ (25%) of the offspring would probably be homozygous for shortness (*tt*)

Because tallness is dominant over shortness, ¾ (75%) of the offspring would probably be tall and ¼ (25%) of the offspring would probably be short.

In the next set of explorations, you will use Punnett squares to predict the possible offspring from parents with the traits you observed earlier in your classmates.

P generation: the parental generation in a breeding.

F1 generation: the first generation of offspring from a breeding.

Exploration 1: Cross two homozygous parents

In *Learning Set 1*, you looked at several traits in your classmates. One trait was attached or detached ear lobes. This trait has two alleles, one for detached ear lobes and one for attached ear lobes. Scientists know that the trait for detached ear lobes is dominant and the trait for attached ear lobes is recessive. Use a capital letter to represent detached ear lobes (*D*) and a lowercase letter to represent attached ear lobes (*d*).

In this exploration, you will cross two homozygous parents, one with attached ear lobes and one with detached ear lobes, to see what kind of ear lobes their children could have.

1. Begin a Punnett square by recording the genotypes of the parents across the top and left side of the square. Scientists call this the **P generation**, or parental generation. Use the diagram to help you. Remember that both parents are homozygous. One has attached ear lobes, and one has detached ear lobes.

2. Now perform the crosses to fill in the four cells in your Punnett square. Scientists call this first generation of offspring the **F1 generation**, or first filial generation. (Filial comes from the Latin word for daughter or son.) Then answer these questions.

a) What are the possible genotypes of all the offspring?

b) What are the possible phenotypes of all the offspring?

c) What percentage of offspring would probably have detached ear lobes?

d) What percentage of offspring would probably have attached ear lobes?

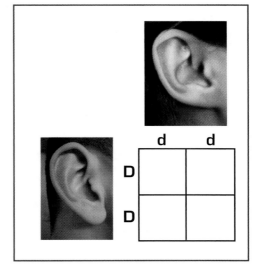

Exploration 2:
Cross a heterozygous parent and a homozygous recessive parent

Another inherited trait in humans is a widow's peak. A widow's peak is a name given to a hairline that comes to a point in the middle. The allele for a widow's peak, *W*, is dominant over the allele for a straight hairline.

Person with a widow's peak. A widow's peak is a dominant trait (Ww or WW).

Procedure

1. Use a Punnett square to show the possible results of a cross between one parent heterozygous for a widow's peak (*Ww*) and one parent homozygous recessive for a widow's peak (with a straight hairline: *ww*). Then answer these questions.

a) What are the possible genotypes of the offspring?

b) What percentage of the offspring of these two parents would probably be homozygous for the dominant allele? What type of hairline will they have?

Person with a straight hairline. A straight hairline is a recessive trait (ww).

GENETICS

c) What percentage would probably be heterozygous? What type of hairline will they have?

d) What percentage would probably be homozygous recessive? What type of hairline will they have?

e) What percentage of the offspring from these two parents would probably have the phenotype of widow's peak?

Exploration 3: Cross a heterozygous parent and a homozygous dominant parent

Blight is a disease of rice caused by bacteria. It is good for rice plants to be resistant to blight. Resistance to blight is dominant. Make the letter *B* stand for resistance to blight. What could be the genotype(s) of a plant resistant to blight?

F2 generation: the second generation of offspring from a breeding.

Procedure

1. Cross two different plants that are resistant to blight. One of the plants is heterozygous, and the other is homozygous. Use a Punnett square to show your results. Then answer the following questions.
    a) What fraction of the offspring would probably be resistant to blight?

    b) What fraction of the offspring would probably be homozygous?

    c) What fraction of the offspring would probably be heterozygous?

2. Next, cross the offspring of the F1 generation. Set up as many Punnett squares as you need for each possible combination of parents. How many do you need? Why?

3. Perform the crosses required to fill in each Punnett square. This is the **F2 generation**, or the second filial generation. Then answer these questions.
    a) What are the possible genotypes of the offspring?

    b) What percentage of the offspring of the F1 generation would probably be homozygous for the dominant allele? Will they be resistant to blight?

    c) What percentage would probably be heterozygous? Will they be resistant to blight?

    d) What percentage would probably be homozygous recessive? Will they be resistant to blight?

e) What percentage of the offspring from the F1 generation would probably have the phenotype of blight resistance?

What's the Point?

From his pea-plant experiments, Mendel discovered that the mathematical principle of probability could be used to predict the outcomes of a genetic cross. Scientists use many tools to study genetics. One genetic tool that they use is the Punnett square, a diagram that predicts the probable genotypes and phenotypes of the offspring of a cross. Often, when designing a procedure for an experiment, information on genotypes and whether a desired trait is dominant or recessive are unknown. In these cases, Punnett squares can be useful in predicting probable outcomes of different crossings. The results of the Punnett squares can then be compared with the actual field data.

Mendel worked with several types of pea plants in his experiments, predicting the outcomes of genetic crossings.

2.5 Plan

How Can Two Varieties of Plants Be Combined to Produce a New Variety?

To: All Student Researchers

From: The Philippine Rice Farmers Cooperative
Quezon Province, Philippines

Subject: Request for Assistance

We are the rice farmers who make up the Philippine Rice Farmers Cooperative
of Quezon Province, the Philippines. We now grow two varieties of rice.
One variety, Type A, has red grains and grows well in many different weather
conditions. The other variety, Type B, has white grains, but the weather has to be
just right for it to grow well.

We would like to grow more rice of the first variety but would like it to produce
white grains, which people prefer, instead of red grains. We do not know the
genotypes of the Type–A rice or the Type–B rice. We do not know which color
is dominant and which is recessive. We are asking student researchers from many
countries to assist us.

If you can tell us how to produce white hybrids of the red rice, we will then
grow the new rice in the field to see which of the hybrids grows best under
different weather conditions. Please send us your recommendations about how
to combine these two varieties of rice. We will be happy to carry out any field
experiments you propose.

Some farmers have sent a request. They need to know how to combine two varieties of rice. One variety produces white grains, which people prefer, but requires just the right weather conditions. The other variety grows better, but it produces red grains, which some people do not like. The farmers want to produce a rice that is white and easy to grow in a variety of weather conditions.

The farmers are asking you to develop a procedure that will allow them to produce a hybrid of the Type-A (red) rice with white grains. They think there may be several ways to produce this kind of hybrid, so they will test each of the white-grain hybrids to see which ones grow well in different weather conditions. You have all the information you need to help them cross the white and red rice. You have explored pollination and fertilization in plants. You know that these processes result in a mixing of the male and female chromosomes and that the new plants that develop from the seeds will show a variety of traits. You also know how to compute the probable genotypes of offspring. However, you do not know yet what the genotypes are of the white and red rice and whether the white color of rice is dominant or recessive. The questions below should help you think through how to figure that out. You will then be ready to design a procedure for the farmers.

Plan

Procedure

1. First, assume that the white trait is dominant and use W to represent the dominant allele.
   - If white rice is dominant, how many genotypes can it have? Why?
   - If white rice is dominant, what will be the genotype or genotypes of red rice?

2. Draw as many Punnett squares as you need to cross white rice (as a dominant allele) and red rice (as a recessive allele). Be sure to label your Punnett squares "white trait dominant."

3. Look at the possible offspring from your Punnett squares. Were any of the offspring white? If you have any white offspring, circle those in your Punnett squares. Save your Punnett squares to compare with the data from the field experiment.

4. Now, assume that the white trait is recessive, and use *w* to represent the recessive allele.

   • If white rice is recessive, how many genotypes can it have? Why?

   • If white rice is recessive, what will be the genotype or genotypes of red rice?

5. Draw as many Punnett squares as you need to cross white rice (as a recessive allele) and red rice (as a dominant allele). Be sure to label your Punnett squares "white trait recessive."

6. Look at the possible offspring from your Punnett squares. Were any of the offspring white? If you have any white offspring, circle those in your Punnett squares. Save your Punnett squares to compare with the data from the field experiment.

Stop and Think

1. Like Mendel did with the peas, the farmers will need to cross-pollinate their plants. In flowering plants, like rice plants, it is necessary to mix the pollen and ovules to produce flowers that form into rice seeds. You might suggest to the farmers to combine Rice A and Rice B. Describe the process they should follow. Will they pollinate Rice A with the pollen of Rice B, or will they pollinate Rice B with the pollen of Rice A or both? Think, too, about how fast the process has to happen. Rice pollen lives for only a few minutes, and rice plants are very difficult to cross-pollinate.

2. Once the first generation of rice plants is grown, you will be able to see the seeds that were produced by your crossings. What will you be able to tell from these seeds? How will you determine if white is dominant or recessive? Describe how it might be possible for the rice to be red but have alleles for white or be white and have alleles for red?

3. How could you use one more generation to get a better idea of the genotypes of each of the seeds you could get from the first generation? Remember to use Punnett squares to try out different ideas and be sure to label them.

Design Your Procedure

You now have ideas about how the farmers might investigate to give you data on developing a hybrid rice with white grains. Using your ideas, begin to design your procedure. The *Rice Field Procedure Planning* page will help you identify all the parts of the procedure you need to think about. Think very carefully about each part. Complete as much of the page as you can. If you need to, go back to your Punnett squares and your answers to the *Stop and Think* questions for help.

Criteria and Constraints

Criteria and constraints are very important because they will help you develop your procedure. Make a table of the criteria and constraints for this investigation on the *Rice Field Procedure Planning* page. Discuss the criteria and constraints with your group. List as many criteria and constraints as you can.

Procedure and Data

The farmers will carry out the procedure you send them. To do this, they will need accurate and detailed instructions. Carefully develop a procedure.

The data the farmers will collect for you will need to be accurate and trustworthy. Be sure you can describe how your procedure will result in accurate data.

Include the following in your procedure:

- how the farmers will carry out the cross-pollination (Be specific in your instructions on how to cross the pollen and ovules.)

- how they should plant the seeds

- what data they should record

- how many generations of rice plants they should grow

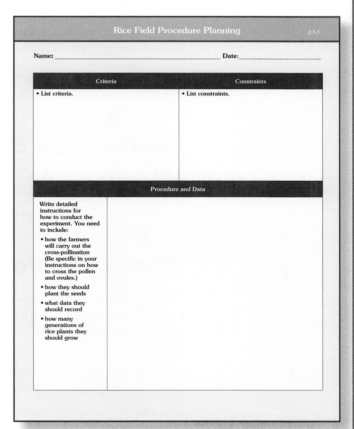

Communicate

Plan Briefing

Others in the class have planned a procedure to answer the same question you are answering, but the class will only send one procedure to the farmers. The purpose of this *Plan Briefing* is to identify which steps are needed in the procedure sent to the farmers and how to tell them what to do. Two groups will present their plans to the class. You will probably see differences and similarities between these plans. In the class discussion, compare the plans to each other and to yours. Identify the strengths of each plan and think about what might need to be improved in each. As each plan is presented, listen for the answers to the following questions:

- What are the steps of the plan? Why did they plan it the way they did? Are there any problems you foresee with this plan?

- What things did they think about before getting to this plan?

- Which parts of this plan belong in the procedure that will be sent to the farmers? What is left out?

As you listen to the presentations, make sure you understand the answers to these questions. If you don't think you have heard the answers, ask questions. Always ask questions respectfully.

Revise Your Plan

With your class, design a procedure you all agree upon for determining how to produce a hybrid white rice from red rice. You will send only one plan to the farmers. This procedure will be sent to the farmers who will carry out your instructions in the field. They will send you the results to analyze later.

What's the Point?

Developing a procedure to produce a new type of organism requires that you think about criteria, constraints, and the phenotypes and genotypes of the parental generation. Sometimes, researchers will not have all the information they need, such as the genotype and whether a trait is dominant or recessive. They can then use Punnett squares to try out the different situations and compare results of the Punnett squares to the actual results from the field. To find out if the traits of an organism are changing, researchers need to observe the organisms' offspring over several generations. When change can be observed and measured over several generations of offspring, the researchers know that the genotype of the population has changed.

2.6 Investigate

Analyze the Results of Your Field Test

The rice farmers ran your procedure in the field and have sent back the results. It is time to analyze the results of the crossings between the two varieties of rice. You will use Punnett squares to do that. Their results are in a memo.

To: All Student Researchers

From: The Philippine Rice Farmers Cooperative,
         Quezon Province, Philippines

Subject: Results of the field experiment you requested

We completed the experiments you suggested and are ready to share the results.
We followed your instructions to cross-pollinate rice plants that have red and white grains. Our goal was to develop a new rice plant of Type–A that has white rice grains.

We started with true-breeding plants. These are ones that have always produced grains the same color as the ones from the parent plant. We know that the trait for rice-grain color comes from a single gene. We also know that a true-breeding plant has the same form of the gene allele for rice-grain color in each of the chromosome pairs in its cells. We do not know which of these colors, red or white, is dominant and which is recessive. To produce the first generation of hybrids, or F1 generation, we crossed the pollen and ovules in two ways:

• We crossed pollen of Type–A rice with ovules of Type–B rice.

• We also crossed ovules of Type–A rice with pollen of Type–B rice.

We discovered that it did not matter which way we crossed the plants.
The phenotypes of the hybrids we produced from the two crossings were the same. We grew four hybrid plants from the crossings. We then observed the color of the rice grains they produced. Here are the results:

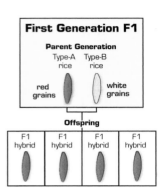

To investigate how the trait for color is passed on to the next generation, we then crossed the hybrids of the F1 generation. We cross-pollinated hybrids in three different ways:

1. We crossed two F1 hybrids.

2. We crossed an F1 hybrid with Type–A rice.

3. We crossed an F1 hybrid with Type–B rice.

The hybrids from each of the three crossings in the second generation, F2, were different. We analyzed the grains of the F2 hybrids for each type of crossing. Here are the results:

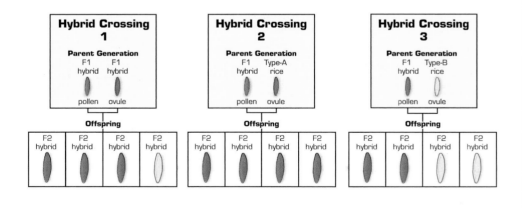

Project-Based Inquiry Science

Analyze Your Data

With your group, use these questions to guide your analysis of the data. Return to your *Rice Field Procedure Planning* page and the class procedure page to remember important parts of the procedure. You will also be using the Punnett Squares you drew earlier to help you develop answers to the questions.

1. The farmers began by pollinating Type–A rice ovules with Type–B pollen and Type–A rice pollen with Type–B rice ovules. What were the results of this part of the experiment? How does that affect further experiments they might do?

2. The farmers grew the four hybrid plants (the F1 generation) by crossing Type-A and Type-B rice, and then they observed the results. Why do you think the rice in this generation is all one color? What do these results tell you about the phenotypes of the seeds?

3. Use a Punnett square to analyze the data from the field experiment. How do the genotypes of the F1 generation compare to the phenotypes? Which genotypes have which phenotypes? Can you determine from this crossing whether the white trait is dominant or recessive? Why or why not?

4. The farmers then crossed the F1 plants to produce another generation. The results are in the F2 boxes. Each crossing is a little different. Use a different Punnett square to show the results of each crossing. Which genotype combinations gave the results shown in the F2 boxes?

5. Compare the Punnett squares you created from the field data with the Punnett squares you created earlier in the planning stage. Which crossings in each produced white rice grains? Which crossings from the field data are the same as crossings from the planning stage? Are these crossings from your "white trait dominant" Punnett squares or from your "white trait recessive" Punnett squares? Using the field data and your Punnett squares, determine whether the trait for white grains is dominant or recessive. Use the F2 generation data and your Punnett squares to support your decision about white grains being dominant or recessive.

6. How do the data from the field experiment help you answer the question: *How can you produce a hybrid of Type–A rice that has white grains?*

Communicate Your Ideas

Investigation Expo

You will share your group's analysis of the field data in an *Investigation Expo*. Each group will prepare a poster to show how you analyzed the field data. Your poster should include the following information:

- Punnett squares that show how the pollen and ovules of the parent plants and hybrids combined to produce the different colors of grains

- the genotypes of the F1 and F2 generations

- the percentage of phenotypes from each of the crossings that resulted in white rice grains

- the percentages of phenotypes from each of the crossings that resulted in red rice grains

- your analysis of which rice-grain color, red or white, is dominant and which is recessive

After the posters are placed on the wall, walk around the classroom and look at the other groups' data analyses. Pay attention to the similarities and differences. As you look at the other posters, make sure you understand how the other groups analyzed the field data. This will help you answer the following questions:

- How are their analyses similar to your analysis and the analyses of others?

- How are their analyses different from your analysis and the analyses of others?

- If the results are different, where do you think those differences came from?

Use evidence from the field data to support your conclusions. Listen carefully as other groups discuss their conclusions, and pay attention to the evidence they present.

Then, as a class, answer the farmers' question: How can they produce a hybrid Type-A rice that has white grains?

Update the *Project Board*

It is now time to update the *Project Board*. You now know more about how traits are passed from plant to plant. Record new science knowledge you gained in Column 3, *What are we learning?* Be sure to add the evidence from the field data in Column 4, *What is our evidence?* You might have more questions now about how traits are passed from one generation to the next. Add ideas for further investigations to the *What do we need to investigate?* column.

What's the Point?

To develop a new plant with a desired trait, it is necessary to cross several generations. The desired trait may not appear in the first generation, but the hybrids still carry the allele for the trait. If the hybrids are crossed in the correct way and enough plants are produced, the trait will show up again in another generation.

2.7 Investigate

How Diverse Are Offspring From the Same Parents?

In *Learning Set 1,* you looked at several traits of your classmates. You probably noticed quite a lot of variation across the class. To help you think more about how traits are transferred from one generation to the next and how this produces variation, you will now investigate variation in a family of Reeze-ots. Working with the Reeze-ots and analyzing the differences in offspring traits will help you see how the mixing of chromosomes from one set of parents can produce genetic variation in offspring.

Observe Parent Reeze-ot Traits

In this investigation, you will observe the amount of variation possible when a pair of organisms with eight traits reproduces. Using the Reeze-ot model, you and your partner will build a pair of Reeze-ot offspring from the same Reeze-ot parents. You will use the same type of lettered trait cards you used the last time you built your Reeze-ots.

You now know a lot more about what the trait cards represent. You know that each card really represents an allele, so the cards could be called allele cards. Each card represents one allele for a trait. The letters on the cards represent the alleles for a trait. Every trait requires two alleles, two different cards with the same letter. Two of the same letter, for example *HH*, *Hh*, or *hh*, describes the genotype for a trait. The uppercase letters on the cards represent the dominant allele and the lowercase letters represent the recessive allele. These letters help you understand the phenotype of the Reeze-ot. The complete set of allele cards for one Reeze-ot includes 16 alleles, two for each trait.

You and your partner will receive two sets of cards. One set represents the traits of the male parent and the other set represents the traits of the female parent. Each pair of students in the class will be using the same sets of cards. They will be building Reeze-ots from the same pair of parents.

Materials

- **1 set of trait cards**
- *Parent and Offspring Reeze-ots Traits page*
- *Key to Reeze-ot Traits page*

Procedure

1. Lay your set of cards on your desk, pairing the allele cards for the same traits. You should end up with eight pairs of cards, one for each of the eight Reeze-ot traits. Copy the genotype for your Reeze-ot into the correct row of the *Parent and Offspring Reeze-ot Traits* page. Copy your partner's parent information into the correct row on the page. Be careful to record the genotypes exactly.

2. Use the *Key to Reeze-ot Traits* page to determine the phenotype for each trait, based on the genotypes you just recorded. Record the phenotype for each trait for the male and female parents. Be careful to record the phenotypes exactly.

Simulate Chromosome Selection

You have recorded all the information for the eight traits for the male and female parent Reeze-ots. The next step is to simulate the production of two Reeze-ot offspring.

Procedure

1. Put your Reeze-ot parent trait cards in a pile. You should have one card pile for one parent and your partner has one pile for the other parent.

2. Without looking at the cards, select one card from your pile and lay it face up on the desk so you and your partner can see the letter. Your partner will select one card from the other parent's pile the same way and lay it on the desk. The letters on these first two cards might not match. Leave room to make the letter matches as you keep drawing cards from

---

Parent and Offspring Reeze-ot Traits 2.7.1

Name: _____ Date: _____

	Height	Leaf color	Leaf number	Number of spikes	Resistance to drought	Resistance to pests	Amount of starch	Number of seeds
Female parent genotype and phenotype								
Male parent genotype and phenotype								
Reeze-ot Offspring-A genotype and phenotype								
Reeze-ot Offspring-B genotype and phenotype								

Key to Reeze-ot Traits 2.1.2 2.7.3

Name: _____ Date: _____

Key to Reeze-ot traits

Trait	Letter Combinations		
Trait H	**HH**	**Hh**	**hh**
Height	Plant is tall. Use 3 marshmallows to build the plant's stem.	Plant is tall. Use 3 marshmallows to build the plant's stem.	Plant is short. Use 1 marshmallow to build the plant's stem.
Trait G	**GG**	**Gg**	**gg**
Color of leaves	Leaves are green. Select the green pipe cleaners.	Leaves are green. Select the green pipe cleaners.	Leaves are white. Select the white pipe cleaners.
Trait L	**LL**	**Ll**	**ll**
Number of leaves	Plant has 2 leaves. Model the leaves using 2 of the pipe cleaners. Insert the leaves into the middle of the stem.	Plant has 2 leaves. Model the leaves using 2 of the pipe cleaners. Insert the leaves into the middle of the stem.	Plant has 1 leaf. Model the leaf using 1 pipe cleaner. Insert the leaf into the middle of the stem.
Trait S	**SS**	**Ss**	**ss**
Number of spikes	Plant has 2 spikes. Insert 2 party toothpicks on top of the stem.	Plant has 2 spikes. Insert 2 party toothpicks on top of the stem.	Plant has 1 spike. Insert 1 party toothpick on top of the stem.
Trait D	**DD**	**Dd**	**dd**
Resistance to drought	Plant is not resistant. No collar.	Plant is not resistant. No collar.	Plant is resistant. Use a piece of green clay to model a collar. Place the collar around the spike, or spikes.
Trait R	**RR**	**Rr**	**rr**
Resistance to pests	Plant is not resistant. No collar.	Plant is not resistant. No collar.	Plant is resistant. Use a piece of yellow clay to model a collar. Place the collar around the spike, or spikes.
Trait Q	**QQ**	**Qq**	**qq**
Amount of starch in seeds	High. Insert a white pushpin in the stem.	Medium. Insert a red pushpin in the stem.	Low Insert a transparent pushpin in the stem.
Trait C	**CC**	**Cc**	**cc**
Number of seeds in each spike	Large. Thread 4 white beads on the spike. Re-insert the spike in the stem.	Medium. Thread 3 white beads on the spike. Re-insert the spike in the stem.	Small. Thread 2 white beads on the spike. Re-insert the spike in the stem.

your pile. After you have selected several cards, you will see that the letters begin to match up. For example, your card for Height (*H* or *h*) should be paired with your partner's card for Height (*H* or *h*).

3. Continue selecting the letter cards, without looking, until both you and your partner have each selected one card for each trait. If, as you are picking your cards, you select a second card for a trait you already have, set the second card aside.

4. You are finished selecting cards when you have eight pairs of allele cards, one pair for each of the eight Reeze-ot traits. Each card pair should include one allele from the female parent and one from the male parent. Check each card pair to make sure it includes one allele from the male parent Reeze-ot and one allele from the female parent Reeze-ot.

5. Record the genotype and phenotype for each trait in the *Parent and Offspring Reeze-ot Traits* page. Add this information to the table in the Offspring-A row.

6. Repeat Steps 1-4 to create the genotype and phenotype for Offspring-B. Record the genotype and phenotype for each trait in the *Parent and Offspring Reeze-ot Traits* page.

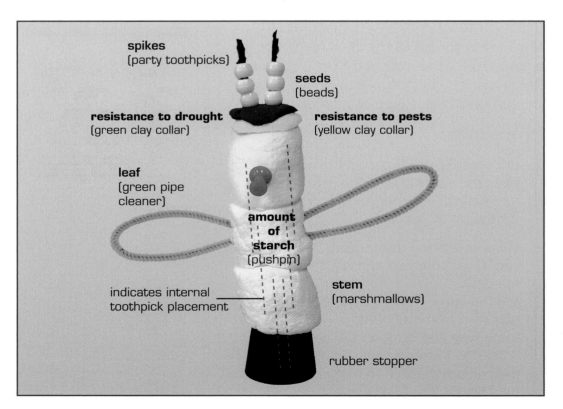

spikes
(party toothpicks)

seeds
(beads)

resistance to drought
(green clay collar)

resistance to pests
(yellow clay collar)

leaf
(green pipe cleaner)

amount of starch
(pushpin)

indicates internal
toothpick placement

stem
(marshmallows)

rubber stopper

Build Your Offspring Reeze-ots

Build both Reeze-ot offspring following the phenotypes you recorded on the *Parent and Offspring Reeze-ot Traits* page. Be sure that you build each offspring using the correct phenotype. Use the same procedure you used when you built the Reeze-ots earlier in the *Learning Set*.

Analyze Your Results

Every member of your class built a different Reeze-ot offspring from the same male and female parent. The Reeze-ot offspring you just built are the first generation of Reeze-ots from these parents. Answer these questions using the Reeze-ots you and your partner just built.

1. How are the phenotypes and genotypes of the two Reeze-ot offspring similar to each other? How are they different? Support your answers with evidence from your *Parent and Offspring Reeze-ot Traits* page.

2. How are the phenotypes and genotypes of the two Reeze-ot offspring similar to the parent Reeze-ots? How are they different? Support your answers with evidence from your *Parent and Offspring Reeze-ot Traits* page.

3. Everyone in your class built one Reeze-ot using the traits of the same pair of parents. How much similarity and difference among Reeze-ot offspring do you think there will be when you look at the Reeze-ots built by your classmates?

Communicate Your Results

Investigation Expo

You built Reeze-ot offspring using the traits selected from one pair of Reeze-ot parents. The goal of the investigation was to observe the possible variations from one set of parents. Because each member of your class randomly selected chromosomes from the same set of parents, you will be able to see some of the variation that is possible.

For this *Investigation Expo*, you will make a poster showing your Reeze-ot offspring. You will describe how the genotype and phenotype of the two Reeze-ot offspring are similar to and different from one another.

By sharing all the Reeze-ots, you will be able to see all of the variation across the offspring.

Begin your poster by drawing diagrams of your Reeze-ots and labeling each trait with the phenotype and genotype information. Be prepared to describe how the genotypes and phenotypes of your offspring are different from and similar to other offspring. Answer these questions in your presentation.

- Which genotypes do each of your Reeze-ot offspring have? How did they get those genotypes?

- Which phenotypes do each of your Reeze-ot offspring have? How did they get those phenotypes?

- What process in plant reproduction did your simulation represent?

- How do your Reeze-ot offspring differ from their parents? How are they similar to their parents?

As other groups present their results, pay attention to the traits other Reeze-ots have and the similarities and differences among the different Reeze-ot offspring. Remember that everyone started with the same pairs of chromosomes. In what ways are you surprised by the results? What might explain the differences among the different Reeze-ots?

Reflect

Now think about the traits the Reeze-ot offspring will have in the next generation and answer the following questions. Be prepared to present your results to the class.

Members of the same family can have some traits in common and can differ on other traits.

1. With your partner, look at the trait your first-generation Reeze-ots (the offspring you built) have for resistance to drought. Using a Punnett square, cross your two first-generation Reeze-ots for this trait. The results of the cross are second-generation Reeze-ots. What are your results? How many of these new Reeze-ots are resistant to drought? How many are not? How do these second-generation Reeze-ots differ in this trait from the first generation?

2. Using Punnett squares, carry out crosses for each of the other seven traits on your first-generation Reeze-ots. What are your results? How do these second-generation Reeze-ots differ in these traits from the first generation?

3. How did the traits change as you used Punnett squares to produce a second generation of Reeze-ots from the first generation? Did you see more variation or less variation? Why?

4. What do you think would happen if you crossed the second generation to produce a third generation? Would you see more or less variation in traits?

Be a Scientist

Modeling and Simulation

Earlier in this Unit, you built a model plant, a Reeze-ot. You built a model plant because you could not build a real plant. A model is used to represent a real object. A simulation is used to represent a real situation. Building models and carrying out simulations are ways scientists re-create the real world in a lab. Scientists try to represent the object or situation they are investigating so it is easy to see important changes over time. They also try to make the model or simulation as accurate as possible. You simulated reproduction in an organism with eight chromosomes and one trait on each chromosome. The model organism was simple, but your simulation of the way traits are inherited was accurate. That let you see how genetic variation becomes greater in each new generation. When you build a model or carry out a simulation, it is always important to think about how it is similar or different from what happens in the real world.

A greenhouse simulates the environmental conditions plants need to grow.

What's the Point?

When a male and a female organism reproduce, half the chromosomes of the offspring come from each parent. When the chromosomes of males and females combine randomly in reproduction, the result is genetic variation in the offspring. There is much variation in the first generation but even more in the second generation.

GENETICS

2.8 Case Study

Genetic Diseases: Cystic Fibrosis

Chromosomes carry many genes. Human chromosomes have over 25,000 genes. All of the information from the organism's genes determines the traits of the organism. Some of the genes carry information that determines common traits such as eye color and height. Some genes carry traits that can cause problems for people. These traits can include genetic diseases.

cystic fibrosis (CF): hereditary disease that causes the body to produce thick, sticky mucus in the lungs, liver, pancreas, and intestines.

mucus: a secretion of the body.

chest physiotherapy: a treatment used for removing the thick mucus that forms in the lungs of a person with CF.

Some genetic diseases can be caused by a dominant allele. If a genetic disease is represented by the letter *A*, a child needs only one allele, the dominant one (genotype *AA* or *Aa*) to have the disease. Only people with the genotype *aa* would not be affected by the disease.

But most genetic diseases in humans are recessive. This means that both the chromosomes a child receives from its parents must contain the recessive allele for the disease. If a genetic disease is represented by the letter *A*, a person without the disease would have the genotype *AA*. A person who carries the disease but does not have the disease would have the genotype *Aa*. A person with the disease would have the genotype *aa*.

Both plants and animals can carry the genes for genetic diseases. Information about genetic diseases is important when developing a new rice plant.

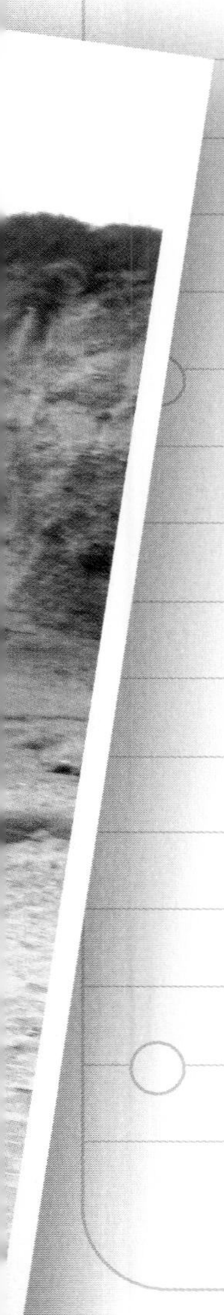

Hello,

My name is Derek. I am 10 years old, and I like to toss the Frisbee high in the sky and catch it. I enjoy playing a lot of games that all kids play. I have **cystic fibrosis** (CF), a disease that affects the lungs and digestive system. Living with cystic fibrosis is one of my greatest challenges. Because of the thick **mucus** in my lungs, I need at least an hour of breathing treatments and **chest physiotherapy** every day. Chest physiotherapy is a treatment used to remove the mucus. To digest food, avoid lung infections, and thin the mucus in my lungs, I take over 30 pills each day. I also get fed overnight with a tube that carries food directly to my stomach. It's a pain. Sometimes people wonder why I cough so much and eat so much. I would destroy CF if I could, but at least all these therapies help me to grow, play, and learn.

Your pal, Derek

From the Desk of L.C.B.

My son, Derek, has cystic fibrosis. We found out he had it when he was only 3 months old. Cystic fibrosis is a genetic condition that affects over 30,000 people in the United States. CF is a recessive trait. One out of every 30 Americans is a carrier of the defective cystic fibrosis allele but does not have any symptoms of the disease. CF results from a **mutation,** a change in the information in a gene. A person having one normal CF allele and one mutated CF allele has no health problems. When a baby has two defective alleles, one from each parent, the child has cystic fibrosis.

People who are affected by CF have a variety of symptoms. These can include difficulty breathing and digesting food. Various therapies treat the complications in the lungs and digestive tract caused by cystic fibrosis. The therapies do not relieve the CF patient of his or her symptoms.

Researchers continue to seek the cure for cystic fibrosis. They have identified the gene that causes cystic fibrosis. It is called the CFTR gene. The CFTR gene is large and complex. For this one gene, there are over 1000 different mutations. As researchers get closer to understanding the gene and how it works, they will be more likely to find a way to cure this disease.

Derek's Mom

Other Genetic Diseases

There are many other genetic diseases in humans. Some are caused by a recessive allele. Some are caused by a dominant allele.

Huntington's disease is a brain disorder that causes uncontrolled movements, mental and emotional problems, and loss of thinking ability. Huntington's disease generally appears when a person is in his or her thirties or forties. By then, the person may have already had children and passed the gene on to the next generation. People with Huntington's disease may have trouble walking, speaking, and swallowing. They may also have changes in personality and trouble with thinking and reasoning. The person with this disease usually survives about 15 to 25 years after symptoms begin.

Sickle-cell anemia affects the blood cells that take oxygen to the body cells. The red blood cells develop a sickle, or crescent, shape. Signs and symptoms of sickle-cell disease begin in early childhood. Children have a low number of healthy red blood cells (**anemia**), infections, and pain. The anemia can cause shortness of breath, tiredness, and slow growth and development. The rapid breakdown of red blood cells may also cause yellowing of the eyes and skin, which are signs of **jaundice**. Pain can occur when sickled red blood cells, which are stiff, get stuck in small blood vessels. Sickle-cell anemia can cause high blood pressure and can lead to heart failure.

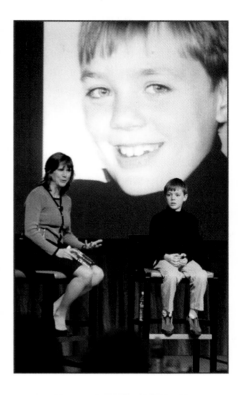

Derek and his mom often speak to large groups about the symptoms and treatments of cystic fibrosis.

mutation: changes to the genetic material of an organism.

Huntington's disease: a fatal genetic disease caused by a dominant allele, which affects the nervous system.

sickle-cell anemia: a genetic disease, carried by a recessive allele, that affects the ability of the blood cells to carry oxygen.

anemia: a low number of red blood cells, which carry oxygen in the blood to the body cells.

jaundice: a condition when pigments from the gall bladder invade the blood. The skin and eyes become yellow.

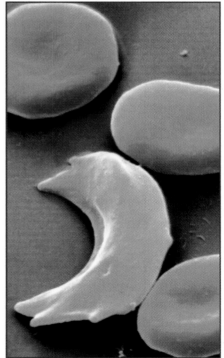

The red blood cells of sickle-cell anemia cannot carry as much oxygen as normal red blood cells.

adult leukemia: cancer of the white blood cells in adults.

Francis Collins

Francis Collins is a leader in the field of genetics. He has spent his life working to find the genetic causes of various diseases. He discovered his love of biology after years of being interested in chemistry. While he was working at the University of Michigan, he became known as a gene hunter. In addition to finding the gene for CF, Dr. Collins found the genes for Huntington's disease and **adult leukemia**, a type of cancer.

Dr. Collins left the University of Michigan in 1993 and became the director of the National Center for Human Genome Research. He is the head of the Human Genome Project. You will learn more about the Human Genome Project later in this Unit.

Reflect

1. Cystic fibrosis (CF) is carried by a recessive allele. A person with CF has two recessive alleles for this trait (cc). Use a Punnett square to determine the percent of offspring that could have CF if parents, who both have the recessive allele (Cc), had offspring.

2. Derek has cystic fibrosis (CF). Even though he has the disease and has to have chest physiotherapy and medication every day, he still does many of the things other kids do. If you met Derek, what would you ask him about living with CF? Would you ask him which sports he likes to play? What other things might you ask him?

Update the *Project Board*

This case study may have made you think more about genes, chromosomes, and how traits are passed from parents to offspring. You may have ideas about new investigations you would like to conduct to explore reproduction and variation. Record your ideas for further investigations in the *What do we need to investigate?* column. The information in this case study may also give you some ideas of how what you have read may be important for answering the *Big Question* or addressing the *Big Challenge*. Record your ideas in the last column on the *Project Board*.

What's the Point?

Genetic diseases can be the result of a dominant or recessive allele on a chromosome. Some genetic diseases can cause people to be very sick. Treatments for genetic diseases are being worked on by genetic researchers.

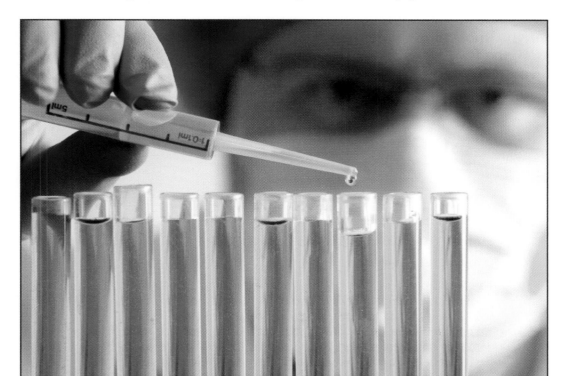

Advances in the field of genetic diseases are made by scientists every day.

Learning Set 2

Back to the Big Challenge

*Make recommendations about developing
a new rice plant that will produce more rice
and more nutritious rice.*

The *Big Challenge* of this Unit is to make recommendations about developing a new rice plant. Because of environmental factors and the needs of a growing population, the recommended rice has to produce more rice seed for eating and more nutritious seeds to provide people with healthier rice. Growing more rice in different conditions and more rice that has greater nutritional value are both important goals for scientists and farmers.

Now that you know more about how traits are passed from generation to generation, the Rice for a Better World Institute thinks you can begin making recommendations.

The Rice for a Better World Institute

To: All Collaborating Scientists

From: The Rice for a Better World Institute (RBWI)

Subject: Research Update

The field procedures on how traits are passed from generation to generation were successful. We are now ready for you to look more closely at the traits we desire in a new rice plant. Earlier, we sent you an inventory of the traits in the rice plants. Using the same experimental procedure you sent us for rice-seed color, we have carried out further experiments. We now know more about how the traits in our inventory are inherited. We discovered that each of these traits comes from a single dominant or recessive allele in a gene. Please look at the table to learn more about these traits.

Rice variety	Trait	Inheritance
A	grows well in dry conditions	recessive
B	grows well even in flood conditions	dominant
C	has high starch content	dominant
D	has high fiber content	recessive
E	has high levels of vitamins and minerals	recessive
F	is resistant to pests	recessive
G	is resistant to disease	recessive
H	produces more rice grains per plant than other rices	recessive
I	requires less fertilizer per acre of rice than other rices	dominant

Please use this new information as you make recommendations about developing a rice plant that produces more nutritious rice, and produces more rice under different weather conditions. We wish you success as you continue your investigations.

Different types of rice have traits that match different criteria of the challenge. Some rice is more nutritious. Other rice types resist pests or diseases. This will result in more growth in the fields. Some of the rice traits are recessive, and some are dominant.

Using what you now understand about rice traits and the transfer of traits from one generation to the next, you will develop three recommendations about how to grow more nutritious rice or more rice.

Using the information in the above table and your new science knowledge from this *Learning Set*, each group will develop a plan for addressing one criterion of the challenge. Groups will then make recommendations about which rice types are most important for the farmers to grow, about which rice types the farmers might combine to produce a variety of rice that has more useful traits, and about how the farmers should combine these plants and the results they could expect to obtain.

Recommend

Using the information sent from the *RBWI*, select traits that will achieve the criterion you are addressing (more rice or more nutritious rice). Read the table carefully and discuss with your group the outcomes of growing rice with particular traits. Think about what would happen to the amount of rice grown if the rice is more resistant to pests or disease. Think about the ways you could make more nutritious rice with more starch or more different types of vitamins and minerals.

You will develop your three recommendations with your group. Use a *Create Your Explanation* page for each recommendation. For each, your recommendation will be your claim. You will add evidence and science knowledge that supports it. Then you will develop a statement linking your recommendation to the evidence and science knowledge.

The first recommendation should focus on which rice types are most important to grow. Focus on the rice trait you are addressing and determine the rice variety, or varieties, you think the farmers should grow. Use the *Table of Traits* to help you make your decisions. Support each recommendation with evidence to help *RBWI* scientists and farmers know why they should trust it. Think about starting your recommendations with *If*, *When*, or *Because*. For example, you might begin a recommendation by writing, "If the farmers wanted to plant seeds that would produce rice that was more nutritious, they should plant seeds that…"

Create Your Explanation

Name:_____ Date:_____

Use this page to explain the lesson of your recent investigations.

Write a brief summary of the results from your investigation. You will use this summary to help you write your Explanation.

Claim – a statement of what you understand or a conclusion that you have reached from an investigation or a set of investigations.

Evidence – data collected during investigations and trends in that data.

Science knowledge – knowledge about how things work. You may have learned this through reading, talking to an expert, discussion, or other experiences.

Write your Explanation using the **Claim**, **Evidence**, and **Science knowledge**.

Your second recommendation should focus on which types of rice you might combine to produce a variety of rice that has traits even more useful for addressing the challenge. Review the various rice types with your group, and think carefully about which rice types you might be able to combine successfully to obtain even better traits. As you write this recommendation, consider beginning it with *Because*. For example, "Because rice would be more nutritious if it had high fiber content and high levels of vitamins and minerals,…" Then complete the statement.

As you work on your recommendations, keep in mind the following big ideas from this *Learning Set*:

- Mendel's experiments with pea plants and inheritance

- how flowering plants reproduce

- self-pollination and cross-pollination

- how recessive and dominant traits are passed on to the next generation

- using Punnett squares to predict what traits the next generation will have

- designing field tests to find out more about traits and how they are inherited

Your third recommendation should suggest to the farmers how they would combine these plants and the results they can expect to obtain. You must support your recommendation with evidence. Your evidence should come from your reading and the Punnett squares to predict the traits the new rice might have. Use one or more Punnett squares to identify the percent of rice that would probably have the important traits. Begin this recommendation with "When farmers use the recommended plants, they should…" Remember to include the expected results of their planting.

Communicate Your Solution

Solution Briefing

After you have developed three recommendations, you will communicate your recommendations to one another in a *Solution Briefing*. To prepare for the briefing, make a poster that includes the criterion you were trying to achieve (more rice or more nutritious rice), the traits your new rice will have, and all of your recommendations.

For each recommendation, be prepared to discuss the evidence and science knowledge that helped you make the recommendation. Make sure to include the important ideas you took into consideration and used. Make sure you also list any questions you still have.

As you listen to your classmates' presentations, make sure you understand their recommendations. If you do not understand something, or if they did not present something clearly enough, ask questions. When you think something can be improved, be sure to contribute your ideas. Be careful to ask your questions and make your suggestions respectfully. As you listen, record your notes on a *Solution-Briefing Notes* page.

Reflect

You have made recommendations about how to develop rice that is either more nutritious or grows better. You have heard the recommendations of other groups for both criteria. Answer these questions in your group, and be prepared to share your ideas with others.

1. How did your recommendations differ from those of other groups who worked with the same criterion? Why did other groups make different recommendations?

2. How were the recommendations similar? What important information did many of the groups use to support their recommendation?

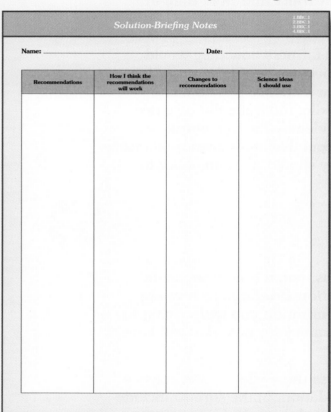

Recommend

As a class, come to an agreement on recommendations for developing a rice plant that combines the two traits you desire: more rice and more nutritious rice.

Update Criteria and Constraints

Now that you know more about achieving the challenge, you may have identified additional criteria and constraints. Or you may feel that the criteria and constraints need to be stated more specifically. Using your new knowledge and evidence from this *Learning Set*, review the list of criteria and constraints. Update the list, making it more accurate. A more accurate list will help you better achieve the challenge.

Update the *Project Board*

The last column on the *Project Board* is the place to record your new knowledge from this *Learning Set* that might help you address the *Big Challenge*—to make recommendations about developing a rice plant that will produce more rice and more nutritious rice. Add your recommendations to the *Project Board* so you can return to them later.

Learning Set 3

How Do Traits and the Environment Interact?

Tall corn plants are grown each year in many parts of the United States. The corn has been selected to produce many ears with plump seeds. But some years, there is not enough rainfall. The corn does not grow as tall or produce as many ears as in more rainy years. Your next challenge is to investigate how the environment and organisms interact, changing how traits are expressed. You will also find out how the environment can cause traits in a population to change over time. Begin by reading the announcement from the research institute. The announcement will provide you with more information about how you can investigate the question for this *Learning Set: How do traits and the environment interact?*

The Canada lynx has long fur that keeps it warm and snowshoe-like feet that allow it to walk on snow.

Research Announcement

To: All Collaborating Scientists

From: The Rice for a Better World Institute (RBWI)

Subject: Research Update

Here at the Rice for a Better World Institute, we thank you for the help you have provided as we try to find a better rice plant for our farmers. Information you have provided has been most useful.

We now have several questions. If the traits we have selected for our new rice plant are passed to every plant in each generation, what will happen if the environment changes? We wonder if the farmers should plant only one type of plant with the traits they desire or if they should plant several different kinds of plants with different traits.

In this next phase of the project, we suggest you investigate whether, over time, it might be best to grow only one type of plant or a mixture of plants with different traits. If you need to run additional investigations, the farmers collaborating on this project will be happy to assist you.

We wish you success as you investigate!

3.1 Understand the Question

Think About the Question

The question for this *Learning Set is How do traits and the environment interact?* You have studied several organisms and their traits. Some traits are more important for survival than others. Some traits can help an individual organism survive when its environment changes. For example, one trait of plants is stem strength. Some plants have very strong stems and other plants have weaker stems. The trait for stem strength varies in the population. In windy areas, a plant with the trait for strong stems will be more likely to survive than a plant with the trait for weak stems. Scientists would say the wind is putting **selection pressure** on the plants. As a result of the selection pressure, the strong-stemmed plants will survive and are more likely to spread their seeds. This increases the chances that more strong-stemmed plants will survive.

People can also put selection pressure on a population. When a farmer pulls up and throws away cabbage plants that aren't big enough to bring to the market, she is selecting for bigger cabbage plants. Her desire for bigger plants to sell is the selection pressure.

Factors in the environment change how a trait is expressed. In order to better understand how organisms will grow, it is important to consider the environment of the organism as well as how a trait exists in a population.

selection: under specific environmental conditions, some traits allow an individual to survive and reproduce better than other traits. The environment (or humans) selects for these traits.

selection pressure: the environment removes organisms with traits that do not help survival and reproduction or keeps organisms with traits that help survival and reproduction.

You will begin exploring the interaction between traits and the environment by designing a model and then using it to run a simulation. In this model, you will use plastic chips in four different colors to represent traits that can vary in organisms in a population. Each color will represent a different trait of an organism that can vary. You will then identify environmental conditions that could affect the growth and survival of organisms with those traits. When you run your simulation, you will remove chips (organisms) from the population when an environmental factor is a problem for your organism.

When the wind blows hard, plants with strong stems survive better than plants with weak stems.

Design Your Model

Procedure

1. With your group, select the organism you will model, and choose four traits for your organism that can very. You can choose to model corn or a bird, or you can choose another organism. Four traits of corn and of birds are already recorded on the *Organisms and Traits* page. If you choose another organism, you will need to choose four traits it might have.

2. For each trait you chose, identify an environmental condition related to that trait that can affect the growth of your organism, and record it on your *Organisms and Traits* page. You can choose factors that are problematic or good for growth of the organism. For example, for corn to grow well, it needs consistent rainfall throughout its growing season. A problematic environmental factor would be too little rain. A good environmental factor would be consistent rainfall.

3. For each environmental factor, record whether that factor will be problematic (decrease the chance for survival and reproduction) or good (increase the chance for survival and reproduction) for the organism.

Materials

- large container of blue, green, yellow, and red chips
- plastic cup per group

Organisms and Traits				3.1.1

Name: _____ Date: _____

	BLUE	GREEN	YELLOW	RED
Corn Traits	Requires consistent rainfall	Has strong stems	Not resistant to pests	Early frost kills seedlings
	Environmental Factors: Heavy rains causes flooding	Environmental Factors: Wind is light	Environmental Factors: Grasshopper population is low	Environmental Factors: Temperatures are rising
Bird Traits	Needs warm temperatures	Nests in trees	Poor fliers	Eats fish
	Environmental Factors:	Environmental Factors:	Environmental Factors:	Environmental Factors:
Organism: Traits				
	Environmental Factors:	Environmental Factors:	Environmental Factors:	Environmental Factors:

Simulate Interactions Between Traits and the Environment

You have prepared your model by selecting an organism, identifying some traits, and determining the possible effects of environmental factors on organisms with various traits. You are now ready to use your model to run your simulation. You will use a *Traits and Trait Frequency* page to record your results.

Procedure

1. Record the four traits of your population on your *Traits and Trait Frequency* page.

2. Take 10 colored chips from the large container—these chips represent your initial population. You must have at least one chip of each color; each color represents a different trait. Carefully select the chips to represent the traits in your population. Discuss your choices with your group. Record the number of chips of each color on your *Traits and Trait Frequency* page. Put the 10 chips in a cup.

3. Simulate environmental factors affecting your population.

   a) Referring to the *Organisms and Traits* page, begin with the trait listed under the first column, Blue. Look at the environmental condition affecting the blue chips. If the condition is problematic for your organism, remove one blue chip from the cup. If it is good for the organism, do nothing. For example, if you are simulating corn and the environmental condition is too much rain, which is problematic for corn crops, you would remove a blue chip from the cup.

   b) Repeat this procedure for each of your chip colors.

4. Repeat Step 3 with the chips remaining in the cup.

5. Empty the remaining chips from your cup and count the number of each color. Record the number of chips of each color on the *Traits and Trait Frequency* page in the row labeled *Final Population*.

Traits and Trait Frequency			3.1.1

Name: _____   Date: _____

Fill in the four traits of your organism from your *Organisms and Traits* page.

Traits for your organism			
Trait 1: Blue	Trait 2: Green	Trait 3: Yellow	Trait 4: Red

Record the frequencies for each chip, or trait.

Initial Population				
Number in Total Population	Frequency of Trait 1 (Blue Chips)	Frequency of Trait 2 (Green Chips)	Frequency of Trait 3 (Yellow Chips)	Frequency of Trait 4 (Red Chips)

Final Population				
Number in Total Population	Frequency of Trait 1 (Blue Chips)	Frequency of Trait 2 (Green Chips)	Frequency of Trait 3 (Yellow Chips)	Frequency of Trait 4 (Red Chips)

frequency (frequencies): the relative number of times a specific value occurs in a total set of data, in this case, the proportion of organisms in a population that has a particular trait.

Analyze Your Data

On your *Traits and Trait Frequency* page you have now recorded the number of organisms with each trait in your initial and final populations. When scientists study populations they look at the **frequency** of various traits—the number of times a trait occurs in the population compared to the total population. The *Calculating Frequencies* box will help you do this calculation.

GENETICS

Be a Scientist

Calculating Frequencies

Frequency is the number of times an item occurs in a set of data compared to the total number of items in the data set. Frequency is calculated by dividing a particular number of items by the total number of items in the set. You will be calculating the frequencies of different traits in a population.

Think about the population of the organisms in your simulation. For example, suppose four of the ten have one particular trait—for example, strong stems. Four of the ten have strong stems and six do not have strong stems. To calculate the frequency of the trait for strong stems in this example, follow these steps:

1. Count the total number of organisms in the population.
   Total number of organisms = 10

2. Count the number of organisms with a particular trait, in this case, strong stems.
   Number of organisms with strong stems = 4

3. Divide the number of organisms with the trait by the total number of organisms in the population. Round this number up or down to one decimal point. This is the frequency of organisms with a particular trait (strong stems) in this population.
   Frequency of organisms with strong stems = $\frac{4}{10}$ *= 0.4*

4. Count the number of organisms without the particular trait.
   Number of organisms without the trait = 6

5. Divide the number of organisms without the particular trait by the total number of organisms in the population. Round this number up or down to one decimal point. This is the frequency of organisms without the trait, strong stems.
   Frequency of organisms without strong stems = $\frac{6}{10}$ *= 0.6*

The sum of the frequencies with and without a trait must always equal 1:
   0.4 + 0.6 = 1.0

In this example, the total number of organisms was 10, so you divided by 10. But your population size will not always be 10. When you calculate frequencies, you must divide by the total number of items in the set.

Analyze your data by first calculating the frequency of traits and then answering the questions below. Be prepared to share your answers with the class.

1. Calculate frequencies of traits.

   a) Calculate frequencies of each color chip in your *Initial Population*, and record them in the part of the chart labeled *Initial Population*. Put a 10 in the first column. That was the number of chips you began with. In this generation, you had 10 chips, so you will divide by 10 when you do the division step.

   b) Calculate frequencies of each color chip in the *Final Population*. Begin by recording the total number of chips you had at the end in the first column of the *Final Population* row of the chart. When you calculate frequencies of these organisms, use this number when you divide.

2. Compare the frequency of traits in the initial population and in the final population. Which chip-color frequencies changed? What forced the frequencies to change?

3 Look at the number of chips you had of each color in the initial and final populations. Look at the frequencies of each color in the initial and final populations. What information do the frequencies give you that you cannot get from the numbers? Why do you think scientists calculate frequencies when they study populations?

4. In this investigation, you were modeling a real-world situation. What were you doing when you removed a colored chip from your cup? What were you doing when you left a chip in the cup?

5. How accurate do you think your model and simulation are? What other factors, in addition to selection pressure, might affect the frequencies of traits in a population?

GENETICS

What Were You Modeling in this Activity?

Your activity with the colored chips simulates what happens in nature. Each chip was a model of an individual organism. The color of the chip represented one trait that could vary. All the colored chips in the cup were a model of a population.

extinct: ceasing to exist.

When you selected a chip based on its color, you simulated the environment placing selection pressure on the population of chips. You selected for one particular color, very much like wind might select for strong or weak stems. Based on the color you selected, you changed the frequency of the phenotypes in the population. The individuals left in the population survived (were left in the cup). Through survival, your population of chips adapted to selection pressure from the environment.

When selection pressures remove individuals from the population, the population decreases in number. Populations can decrease in number rapidly when the environmental conditions change suddenly. Some populations decrease to such a small number that they cease to exist any longer. These populations become **extinct**. The selection pressures that cause populations to change in number can come from the environment, such as wind, or from people, such as selection by farmers for the size of cabbages.

Communicate Your Results

Investigation Expo

Each group in your class ran a different model. You will present your results to the class so that others will be able to observe and analyze your results. To prepare for your presentation, create a poster to share with the class. Include the following things on your poster:

- Describe your organism and its traits.

- Describe the environmental factors you chose and your decision about whether that factor harms the organism or has no effect.

- Describe how you determined the frequencies for your Initial and Final Populations.

- Show all the data you collected.

- Describe how and why the population changed over time.

- Describe why scientist calculate frequencies.

- Identify the environmental effect (selection pressure) that changed your population trait frequencies the most.

As you are listening to each presentation, think about the environmental factors each group simulated and how you think each factor might affect the population. Do you think the group made a good decision about removing or leaving a chip in the cup? What might you have done differently if you had been building the model? Discuss as a class how selection for a particular trait might affect the frequency of traits in a population.

Explain

This simulation activity probably gave you many insights about how populations change over time. Use your own group data, the data of other groups, and discussions you had after the *Investigation Expo* to develop an explanation of the way the environment affects the traits of organisms in a population. Use a *Create Your Explanation* page to help you develop your explanation, and support it with evidence from your model. The *Create Your Explanation* page helps you make sure the explanation takes into account your claim, evidence, and science knowledge. Begin by writing a claim about the change of the color frequencies in the models. The results from the chip investigations are your evidence. Record the science knowledge that helped you consider the environmental factors. Write a statement that pulls together your evidence and science knowledge to support your claim. Use *because* in your statement. Your statement should help others understand what is happening in nature that makes your claim valid. Remember that an explanation is the best you can do given what you know. You will get a chance to revise this claim and its explanation later in the *Learning Set*.

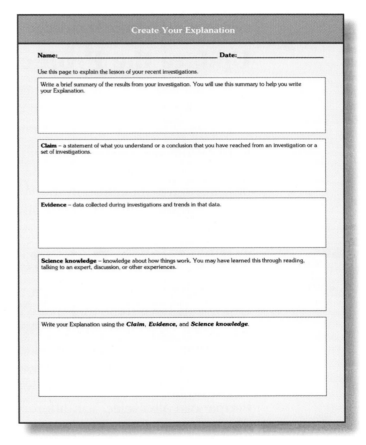

Create Your Explanation

Name:_____ Date:_____

Use this page to explain the lesson of your recent investigations.

Write a brief summary of the results from your investigation. You will use this summary to help you write your Explanation.

Claim – a statement of what you understand or a conclusion that you have reached from an investigation or a set of investigations.

Evidence – data collected during investigations and trends in that data.

Science knowledge – knowledge about how things work. You may have learned this through reading, talking to an expert, discussion, or other experiences.

Write your Explanation using the *Claim, Evidence,* and *Science knowledge*.

Communicate

Share Your Explanation

Share your group's explanation with the class. When you share your explanation, tell the class what makes your claim accurate based on your evidence and science knowledge. Report, too, on what else you think you need to know to make a better claim or support your claim better. As each group shares its explanation, pay special attention to how the other groups have supported their claims with evidence and science knowledge. Ask questions or make suggestions if you think a group's claim is not as accurate as it could be or if the group has not supported its claim well enough with evidence and science knowledge.

Update the *Project Board*

In this activity, you simulated selection pressure on a population. You saw that the frequencies of traits can change in a population because of selection pressure. What caused the frequencies of traits to change over generations?

In the *What do we think we know?* column, record what the simulation showed you about the role of the environment in changing the frequency of traits in a population. In the *What do we need to investigate?* column, record questions you still have about the way the environment can change a population.

What's the Point?

Pressures from the environment select specific traits in a population. Organisms have traits that make them more or less likely to survive and reproduce. Selection pressure changes the frequency of different traits in the population. One example of selection by the environment might be wind selecting for strong or weak stems in a population of plants. If only plants with strong stems are left, they will pass that trait on to the next generation. People can also select for different traits. People selecting large cabbages over small ones is an example of people affecting a population through selection. When the environment or people select for specific traits, they are putting selection pressure on the population. Extreme selection pressures can cause a population to become extinct.

3.2 Explore

How Do Animal and Plant Populations Change Because of the Environment?

The polar bear has thick hair to keep it warm.

Organisms have a variety of traits that make them more or less likely to survive and reproduce. As you saw with the chips, selection pressure on a population can come from the environment and ultimately cause the frequency of traits in a population to change. Organisms with the traits (phenotype) best suited for an environment survive and reproduce. Because these organisms can survive and reproduce, the frequency of their traits increases. In this section, you will further explore how selection pressure can make that happen as you investigate the question, *How do animal and plant populations change because of the environment?*

When the frequency of traits in a population changes because of selection pressure, the population is adapting to the environment. An **adaptation** is an inherited trait or set of traits that improves an organism's chance of survival and reproduction in its environment.

Organisms living in particular environments might always seem to have the exact traits they need to survive and reproduce. How do you think organisms have become so well adapted to their environment? Keep this question in mind as you explore adaptations and think about how they help organisms survive and reproduce.

Kinds of Adaptations

Adaptations can help organisms survive in their enviroment. They can help an organism hide and keep from being eaten. They can help an organism catch food to eat. They can protect the organism from **predators**. Most animals are **prey** for other animals. There are two types of adaptations: **structural adaptations** and **behavioral adaptations**.

Structural adaptations include body shape and size, structure of legs and feet, body covering, and types of ears and their placement on the head. Behavioral adaptations are important because they cause an organism to act in predictable ways. Warning calls of animals, an animal's response to attacks,

adaptation: an inherited trait or set of traits that improve an organism's chance of survival and reproduction in its environment.

predator: an animal that eats other animals.

prey: animals that are eaten by other animals.

structural adaptation: adaptations that are part of an organism's physical makeup.

behavioral adaptations: adaptations that cause an organism to act in a specific way.

and the feeding behaviors of organisms are all examples of behavioral adaptations.

Organisms often have both structural and behavioral adaptations to make them better suited for survival and reproduction in their environment. For example, snowy owls in the arctic **tundra** have thick feathers that hold warm air near their bodies. During the long arctic winter, the snowy owl can fly south where there is more food. Flying south is a behavioral adaptation. The thick feathers are a structural adaptation.

The thick layer of feathers on the snowy owl holds a lot of air near its body and keeps it warm.

tundra: a rolling, treeless plain found in far northern areas of Earth.

species: a group of organisms whose members have the same structural traits and can breed with one another.

Butterfly Adaptations

The butterflies below look very much alike, but they are different **species**. The Monarch butterfly, on the left, stores bad-tasting chemicals in its body. When a predator eats a Monarch, it swallows the bad chemicals, becomes sick, and learns to leave other Monarchs alone. The Viceroy butterfly, on the right, looks very much like the Monarch butterfly. It also has a bitter taste, though not as bitter as the Monarch.

Monarch butterfly

Viceroy butterfly

1. The Monarch butterfly has the structural adaptation of being brightly colored. Why might this be an advantage for this butterfly?

2. Compare the appearance of the Monarch and Viceroy butterflies. How would the Viceroy's coloration be an advantage for its survival?

3. Even though it is brightly colored and bad tasting, the Monarch butterfly does get eaten sometimes. Its bright colors and bitter taste do not protect it every time. Why do you think this is so?

Stop and Think

1. What happens when some individuals in a population have traits that make them better adapted to the environment? Give an example to support your answer.

2. Using your investigation and reading experiences, describe how a population, as a whole, becomes better adapted to the environment over time.

Explore

How Does Selection Pressure Change the Frequency of the Trait Bug Hunt Speed in a Population?

Now that you know more about how adaptations help individuals survive and reproduce, you will use NetLogo to explore how the frequency of an adaptation in a model population changes because of selection pressure. One environmental factor that can affect the frequency of traits in a population is selection pressure from a predator. When a predator selects specific individuals from a population, the selection pressure can cause the population to adapt.

An example of this is the relationship between birds and **bugs**. Birds are predators that eat bugs for energy. The more bugs the bird eats, the more energy it gains from food. However, the harder the bird works to get its food, the more energy it uses. Every time the bird moves, it uses energy. To survive, the bird needs to find the most food while using the least amount of energy.

bug: an insect with sucking mouthparts.

The bug also needs to survive. The bird places selection pressure on the bugs. This selection pressure will, over time, change the population of bugs.

Design a Population Model

Imagine a model environment that consists of a large dirt field where a population of bugs lives. In this same field there is also a bird that eats the bugs.

What you need to know about the bugs in this model environment:

- Bugs move around, searching for food. Some bugs are faster than others. The speed at which a bug moves is an inherited trait.

- Bugs move in straight lines.

- Bugs do not try to run from the bird or to avoid it in any way.

- Every time the bird eats a bug, a new bug is born from the remaining

GENETICS

population of bugs. The population of bugs always remains the same size, but the number of faster bugs and slower bugs changes, depending on what kind of bug the bird eats.

What you need to know about the bird in this model environment:

- Every time a bird moves, it uses energy. Every time it eats a bug, it gains energy.

- The bird can use three hunting **strategies**, or behavioral adaptations, that improve its survival.

Sit and Wait: The bird sits in one spot and waits for bugs to come to it. When a bug bumps into the bird, the bird eats it.

Chase Prey: The bird chases after bugs, attempting to catch as many as it can.

Move Randomly: The bird moves around the field randomly and hopes it bumps into bugs.

strategy (strategies): a way of doing something that gives an organism an advantage over other organisms.

parameters: the limits or boundaries (in this case, values).

line graph: a type of graph that uses points plotted and then connected to form a line.

Be a Scientist

Using NetLogo

To understand how a predator can affect a prey population, you will use a computer program called NetLogo. You will simulate interactions between a predator (bird) and its prey (bugs). When the bird selects specific bugs, you will observe how the bug population adapts. Your group will simulate a bird using each hunting strategy on a population of bugs.

First, open the program on your computer. Your teacher will tell you which model to open. Once the model is loaded, you should see a screen like the one shown.

The Model Set-up Window is on the left. In this window, you will select the beginning **parameters** of your simulation.

In the center at the very top is a *Speed Slider* button. This slide button speeds up or slows down the bugs relative to one another. Fast bugs will still move faster, and slow bugs will still move slower, but the entire population will speed up or slow down.

The rest of the screen is divided into three main sections:

- The *model input settings* are on the top left.
- The *plot graph windows* are on the bottom left.
- The *graphics window* is on the right.

The *model input settings* are where you set the parameters for your initial population of bugs. This area consists of four parts:

- *Initial-bugs-each-speed*. This slider button controls the number of bugs that begin the model. The bugs are randomly distributed around the field. The computer assigns each bug a speed. There are six different speeds. Sliding the red bar controls the number of bugs your model field begins with at each speed. For the number of bugs you choose, you will have 1 bug at each speed. The overall population of bugs is determined by multiplying this value by 6. For example, if you choose *2*, you will have 2 bugs each at 6 different speeds in the initial population, or 12 bugs.

- *Alive bugs*. This display shows the total number of bugs in the population. For every bug eaten, another will be born at random from the remaining bugs. For example, if you set your *Initial-bugs-each-speed* at *2*, you will always have 12 bugs. This number will remain the same throughout the simulation. However, as your predator eats bugs, the number of fast bugs and slow bugs changes.

- *Total caught*. This display shows how many total bugs your predator has eaten. This number will change throughout the simulation.

- *Speed-color-map*. This button controls the colors of your bugs. Bugs that move at different speeds will be a different color. The lighter the color, the faster the bug moves.

You will be using two graphs in the *plot graph windows* area: *Average bug speed vs. time*, a **line graph**, and *Frequency of bugs*, a **bar graph**, a type of graph that uses columns to show frequencies. By analyzing the data on these two graphs, you will observe what happens to your bug population under selection pressure by the bird.

- *Average bug speed vs. time*: If the population has more fast bugs than slow bugs as a result of selection pressure, the average bug speed will increase. If the population has more slow bugs than fast bugs as a result of selection pressure by the predator, the average bug speed will decrease.

- *Frequency of bugs*: If the population has more fast bugs than slow bugs as a result of selection pressure, the line plot will shift to the right. If the population has more slow bugs than fast bugs as a result of selection pressure by the predator, the bar graph will shift to the left.

Simulate *Sit and Wait*

In this strategy, the bird stays in one place and waits for bugs to bump into it.

Predict

Predict the effects on the bug population for the *sit and wait* strategy. Record your prediction of how and why you think the bug population will change. Use the characteristics of the bugs and bird to help you.

Run a Selection Pressure Simulation

In this simulation, you will observe how the *sit and wait* strategy affects the trait of bug speed in the population of bugs. As you make your observations, keep your prediction in mind.

Procedure

1. First, move the slider on your *Initial-bug-each-speed* to the right to *10* to set your initial population of bugs at ten bugs for each speed. Your initial population will consist of 60 bugs.

2. Set the *Speed-color-map* to *violet shades*. Remember, the lighter a bug, the faster it moves.

3. Move the *Speed Slider Button* to the right to the middle position, normal speed.

4. Press the *setup* button. This button will set your initial parameters into the field.

5. The initial conditions of your selection pressure model are already recorded in your *Model Selection Pressure Observations* page. Look at the bottom left of the *plot graph windows* and draw the line plot for the *Frequency of bugs* in the box on your *Model Selection Pressure Observations* page. This graph shows your beginning population. The *Average bug speed vs. time* window will be empty until the simulation begins.

6. Press the GO button and move your cursor to the *graphics window*. Choose a spot on the field, and hold down the mouse button. Remain in this position for 90 seconds without moving the cursor around.

7. After 90 seconds, release the mouse button and press the GO button to stop the simulation.

8. Record the number of bugs eaten on your *Model Selection Pressure Observations* page. Draw the *Frequency of bugs* and *Average bug speed vs. time* graphs. These graphs represent your ending population.

Simulate *Chase Prey*

In this strategy, the bird selects a bug to eat and goes after it.

Predict

For the *chase prey* strategy, predict the effect on the bug population. Record your predictions of how and why you think the bug population will change. Use the characteristics of the bugs and bird to help you.

Run a Selection Pressure Simulation

In this simulation, you will observe how the *chase prey* strategy affects the trait of bug speed in the population of bugs. As you make your observations, keep in mind your predictions.

Procedure

1. First, move the slider on your *Initial-bug-each-speed* to the right to *10* to set your initial population of bugs at ten bugs for each speed. Your initial population will consist of 60 bugs.

2. Set the *Speed-color-map* to *violet shades*. Remember, the lighter a bug, the faster it moves.

3. Move the *Speed Slider* button to the right to the middle position, normal speed.

4. Press the SETUP button. This button will set your initial parameters into the field.

5. The initial conditions of your selection pressure model are already recorded in your *Model Selection Pressure Observations* page. Look at the bottom left of the *plot graph windows* and draw the bar graph for the *Frequency of bugs* in the box on your *Model Selection Pressure Observations* page. This graph shows your beginning population. The *Average bug speed vs. time* window will be empty until the simulation begins.

Model Selection Pressure Observations

Name: _____ Date: _____

Initial settings			
Initial-bug-each-speed – **10**	Total population – **60**	Speed-color-map – **violet**	Speed slider button – **normal**

Initial *Frequency of bugs* bar graph

1ˢᵗ Strategy: sit and wait	Total bugs eaten:
Frequency of bugs bar graph	*Average bug speed vs. time* line graph

2ⁿᵈ Strategy: chase prey	Total bugs eaten:
Frequency of bugs bar graph	*Average bug speed vs. time* line graph

3ʳᵈ Strategy: move randomly	Total bugs eaten:
Frequency of bugs bar graph	*Average bug speed vs. time* line graph

GENETICS

6. Press the GO button and move your cursor to the *graphics window*. Move the mouse around the window to chase the prey, and click the mouse when you are on top of a bug. Continue to chase and eat bugs for 90 seconds.

7. After 90 seconds, press the GO button to stop the simulation.

8. Record the number of bugs eaten on your *Model Selection Pressure Observations* page. Draw the *Frequency of bugs* and *Average bug speed vs. time* graphs. These graphs represent your ending population.

Simulate *Move Randomly*

In this strategy, the bird moves around without a pattern. The bird eats a bug when either it bumps into a bug or a bug bumps into it.

Predict

Predict the effect on the bug population of the *move randomly* strategy. Record your predictions of how and why you think the bug population will change. Use the characteristics of the bugs and the bird to help you.

Run a Selection Pressure Simulation

In this simulation, you will observe how the *move randomly* strategy affects the trait of bug speed in the population of bugs. As you make your observations, keep in mind your predictions.

Procedure

1. First, move the slider on your *Initial-bug-each-speed* to the right to *10* to set your initial population of bugs at ten bugs for each speed. Your initial population will consist of 60 bugs.

2. Set the *Speed-color-map* to *violet shades*. Remember, the lighter a bug, the faster it moves.

3. Move the *Speed Slider Button* to the right to the middle position, normal speed.

4. Press the *setup* button. This button will set your initial parameters into the field.

5. The initial conditions of your selection pressure model are already recorded in your *Model Selection Pressure Observations* page. Look at the bottom left of the *plot graph windows* and draw the bar graph for the *Frequency*

of bugs in the box on your *Model Selection Pressure Observations* page. This graph shows your beginning population. The *Average bug speed vs. time* window will be empty until the simulation begins.

6. Press the *go* button and move your cursor to the *graphics window*. Hold down the mouse button and move the cursor around the field randomly. Do not try to chase any of the bugs or remain still for more than a second or two. Continue to move randomly for 90 seconds.

7. After 90 seconds, release the mouse button and press the *go* button to stop the simulation.

8. Record the number of bugs eaten on your *Model Selection Pressure Observations* page. Draw the *Frequency of bugs* and *Average bug speed vs. time* graphs. These graphs represent your ending population.

Analyze Your Data

In your group, discuss the data you collected from your simulations and answer the following questions. Be prepared to share your answers with the class.

- Compare the graphs of your beginning and ending bug populations. The same number of bugs existed at each speed at the beginning of each simulation. How do the graphs look after running the simulations? How did the graphs change?

- What changed on the *Average bug speed vs. time* graph as your bird used each strategy? What does that show?

- With each strategy, was your bird better able to catch faster bugs or slower bugs? How did the number of bugs at each speed change as your bird ate bugs?

Communicate Your Results

Investigation Expo

Make a poster to present the results of your simulations. Include the following on your poster:

- The graphs from your *Model Selection Pressure Observations* pages.

- Your answers to the Analyze Your Data questions.

- Your answers to these questions:

    - How did the different kinds of selection pressure on the bug population cause the traits of each population to change?

    - Why do you think bug speed changes differently when under selection pressure by different predator strategies?

    - How did what you observed compare with your predictions? How did what others observed compare with your predictions?

As you hear from the other groups, compare their results to yours. Were the averages of your group different from those of the other groups? Were averages of your group different from the averages of the whole class?

Reflect

1. With which strategy did the bird catch the most bugs? Why was this strategy successful in catching more bugs? How successful do you think this strategy will be over a long period of time? What evidence do you have to support your answer?

2. With which strategy did the predator catch the fewest bugs? Why was this strategy not as successful in catching a large number of bugs? What advantage does this strategy have that the other strategies do not have? Is it more likely to be successful over a longer period of time? Why or why not?

3. Assume that the predator can use only one strategy, *chasing prey*. What is likely to happen to the bug population in the future? What is likely to happen to the predator in the future? In this situation, which of the organisms adapted?

4. Think about a field of corn. Bugs are pests that eat the corn. Birds are predators that eat the bugs. How is the simulation you carried out similar to what happens in a real corn field? How is it different?

Expand Your Thinking to Rice

Think about a rice field with pests that feed on the rice. It is important to stop the bugs that are eating the rice. To do this, you introduce predators of those bugs into the field. Think about how the computer simulation might have given you information about this problem. Knowing what you now know about selection pressure and how populations adapt, answer the following questions:

- What criteria and constraints would you have to consider before you introduced the predators?

- What traits would you want the predators to have?

- What predator hunting strategies would be most useful? Would you introduce a predator with only one hunting strategy or would you introduce more than one kind of predator, each with a different hunting strategy? Why? Use evidence from your simulation to justify your answers.

- What might be the dangers of introducing many different kinds of predators into the environment?

What's the Point?

Selection pressure placed on a population causes changes in that population. Changes in traits of a population that occur as a result of selection pressure are called adaptations. Adaptations are inherited traits or sets of traits that improve an organism's chance of survival and reproduction in its environment. Structural and behavioral adaptations are both important for survival and reproduction. Organisms that can adapt to changes in the environment will be more likely to survive and reproduce.

3.3 Read

How Do Scientists Know About the Effects of Selection Pressure?

naturalist:
a person who studies the plants, animals, and environment of an area.

Charles Darwin was a **naturalist**, a person who studies the plants, animals, and environment of an area. He lived in the 19th century and came from a wealthy family. Darwin was passionate about natural history and often observed the characteristics of animals and the world around him.

As a young man, he signed on to be the naturalist aboard the ship, the HMS Beagle. The ship was scheduled to spend five years mapping the coastline of South America. On the Galapagos Islands, a cluster of volcanic islands off the coast of Ecuador, Darwin saw many different species, groups of organisms of the same kind with the same structural traits. However, on different islands, animals that seemed to be the same species varied greatly in size, shape, and color. One of the animals he observed was a small bird, a finch. The finches on the different islands had many different beak shapes. Darwin was surprised by this observation and he asked this question: Why does the beak shape in finches vary so much? He recorded observations about birds and other animals and their traits and took his notes back to England.

The Galapagos Islands are a cluster of islands 600 miles west of Ecuador, South America.

Charles Darwin went on a science expedition around the world on the HMS Beagle. He studied plants and animals everywhere he went. He also collected specimens for further study.

Ecuador

Galapagos Is.

Darwin observed many species found only on the Galapagos Islands, such as the enormous Galapagos tortoises.

According to Darwin's observations, finches on the different Galapagos Islands had a variety of beak shapes.

Darwin gathered all his evidence about animals' traits from his observations. He made sure he had recorded his observations accurately. He then used his notes, his observations, and his knowledge of geology to develop a **hypothesis** about how he thought the animals had gotten their different traits. A hypothesis is a prediction based on observations. It predicts what will happen between an independent (manipulated) variable and a dependent (responding) variable. Using the evidence from his observations, and what he had learned from other scientists, Darwin developed a theory explaining the variation. A scientific theory is a broad explanation that is strongly supported by a body of evidence.

Darwin lived in a time before anyone knew about heredity—how traits are passed down from generation to generation. Darwin was an educated man, but he knew nothing about chromosomes or genes. Most people at this time thought all species originated about 6000 years ago and remained unchanged through history. Darwin's interest in natural history came from his study of the work of other scientists, especially **geologists**, scientists who study the origin, history, and structure of Earth. Darwin knew about **fossils**, the remains of plants and animals preserved in rocks in Earth's crust. Looking at fossils made him wonder how species change over time.

On November 24, 1859, nearly 25 years after his voyage to the Galapagos, Darwin's book, *On the Origin of Species*, was published. In the book, he described his observations, predictions, and theories. His theories were supported by the evidence he had collected in the Galapagos. Darwin questioned why some individuals in a population die early but others do

hypothesis (pl, hypotheses): a prediction of what will happen between an independent (manipulated) variable and a dependent (responding) variable.

geologist: a scientist who studies the origin, history, and structure of Earth.

fossils: the remains of once-living organisms preserved in rocks in Earth's crust.

not, and why some species survive and others do not. From his study of fossils, he knew some species that lived in the past no longer existed—they had become extinct. He thought there must be a struggle for survival. He thought the individuals of a species must compete for the things they need to survive. These needs include food, light, water, and places to live. According to his theory, individuals with the *most well-suited* traits will survive and produce offspring more often than individuals without them.

Using his evidence, Darwin also predicted that traits that make an organism able to survive in its environment are passed to the next generation. This happened, he explained, because those organisms that survive will pass on those traits to their offspring. This way, over many generations, the number of individuals with the preferred traits will increase, and the number with the less beneficial traits will decrease. In extreme cases, where organisms cannot adapt to the selection pressure, species could become extinct.

Darwin called this process of survival and reproduction **natural selection**. He thought several factors were involved in natural selection.

natural selection: the differences in survival and reproduction among members of a population as a result of selection pressure.

Drought conditions will affect plants and animals. Native food sources for animals, both meat eaters and plant eaters, may not be available. Only the plants and animals able to survive with little or no water will survive.

- There is variation among individuals in a population.

- Traits are passed from parents to offspring.

- The size of populations tends to increase.

- There is a *struggle for survival* (or competition for the needs of life) among individuals.

- Based on the traits they have, some individuals in a population survive and reproduce better than others.

- There is a difference in the traits passed on to the next generations.

Over 150 years ago, Darwin's book, *On the Origin of Species,* transformed how biologists view changes in traits and populations. Many scientists have built upon Darwin's original hypotheses. They have observed many other species and have applied Darwin's ideas to changes in other species. The theory of natural selection has been used to explain a variety of changes in many different organisms.

Most recently, Darwin's ideas on how new species come about have been confirmed by the discovery of chromosomes and genes. Scientists now know that chromosomes and genes are passed to the next generation, and that changes in chromosomes and genes are also passed to the next generation. Scientists have looked at the genes of many different organisms and have learned that many species have genes in common. The more closely species are related, the more similar their genes are.

Alfred Russel Wallace

Sometimes in science more than one scientist develops the same theory at about the same time. Albert Russel Wallace was also a scientist who had sailed to South America to collect plants and animals. He had read the same books Darwin had read. His ideas agreed with Darwin's in many ways. Before Darwin published *On the Origin of Species*, Wallace published four papers on his observations. Wallace used the same term, "natural selection," to explain his theory. When Wallace sent his fourth paper to Darwin, Darwin rushed to finish his writing. Although Wallace had published first, Darwin's work gained more fame. But Darwin and Wallace became friends and remained friends throughout their lives.

Stop and Think

1. How did Darwin's experiences help him to understand how species change and how new species develop? How did he use these experiences to come up with his theory of natural selection?

2. What did Darwin see as the major factors causing change from generation to generation?

Evolution and Natural Selection

In the bug-speed simulation, the bug population changed due to selection pressure from the bird. Bugs with traits that made them easier to catch were removed from the population. This process is natural selection. Individuals were eaten, based on these traits, and that changed the frequency of those traits in the population. Scientists call the result of that change **evolution**. Although evolution and natural selection are related, they are not the same thing. Evolution is a change in the frequencies of traits in a population. Natural selection causes populations to change how they look or behave. These changes are brought about by selection pressure. Populations evolve, but individuals do not.

evolution: the change in the frequencies of traits in a population over time.

In responding to selection pressure, many species of animals, such as these camels, adapted to life in a desert environment.

All populations adapt to their environment. They do not necessarily become *better* over a long period of time. As you saw in the bug-speed simulation, when bugs adapted to one strategy of the bird, the bird could change its strategy. A trait or strategy that is successful at one time may be unsuccessful at another. A trait that is an advantage at one time may be a disadvantage when the environment changes. Any change in the frequency of traits in a population is evolution. Evolution is the result of change; it is not the process of change.

When Darwin proposed his theory of natural selection, he knew nothing about evolution. The word was not used at that time. Natural selection is Darwin's theory that explains *how* species and populations change over time. Other scientists after Darwin studied the results of natural selection. They concluded that selection pressure often results in a population becoming better adapted to its environment. They called the theory that explains the results of the change evolution.

Update the *Project Board*

Now it is time to update your *Project Board*. You now know more about adaptation, natural selection, and evolution. Record what you now know in the *What are we learning?* column. If you have questions or ideas for investigations, record them in the *What do we need to investigate?* column. As you are listing your questions, think about how the environment can change the traits of a population. Also, if you have recommendations related to the *Big Question*, add that to the last column.

What's the Point?

Charles Darwin studied variations in traits in animal populations and wondered how individuals got different traits. Using his observations, he developed a theory of natural selection. His theory explained how individuals with the *best* traits for the environment would live and reproduce better than those without these traits. Darwin's theory also explained how the species he observed came from ancestors that lived long ago. Through the process of natural selection, one species developed into the many species he observed. Darwin published his observations and theory in a book, *On the Origin of Species*.

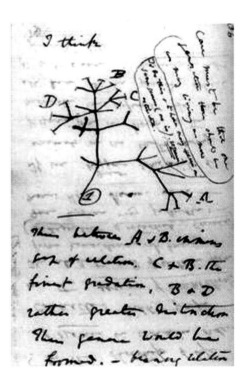

Tree of Life: the first known sketch by Charles Darwin of an evolutionary tree describing the relationships among groups of organisms.

Scientists continued to study Darwin's data and to read his book. They derived a theory about the results of natural selection. That theory is called evolution. Natural selection is a process of struggle for existence that explains variation, survival of the fittest, and adaptation. Evolution is the result of that process.

GENETICS

More to Learn

Fossil Evidence for Evolution

Charles Darwin used the work of other scientists to develop his theory of natural selection. He read papers written by geologists and saw that geologists were often finding evidence of change within species and increasing complexity in plants and animals over time. These findings helped Darwin realize that species had changed before and were probably still changing.

In the 1800s, as today, geologists relied on fossils as clues to what life was like in the past. Each fossil is like a *snapshot* of a plant or animal at a particular time in Earth's history. By sequencing fossils according to when they formed, scientists can understand how living things have changed, or evolved, over time. Fossils also provide a snapshot of what kinds of environments existed millions of years ago.

Over time, **paleontologists** have carefully studied a large number of fossils. They have made two important observations:

paleontologists: scientists who study fossils to learn about living things that existed in the past.

Frozen remains of a nearly intact woolly mammoth discovered in Siberia in 1997. The mammoth lay atop clay soil filled with frozen prehistoric plants that still had their original green color. A blue flag marks the find.

First, paleontologists now know that species have changed over time. This is what Darwin's theory predicted. For example, 60 million-year-old fossils, very similar to small horses, have been found. Paleontologists believe these small mammals were the ancestors of today's horses. As scientists examine more and more recent fossils of horse-like mammals, they observe gradual changes in structure. Paleontologists believe that horses changed as the environment changed. Horses with characteristics that helped them survive in the new environments survived, while those lacking such characteristics became extinct. This is one example of how the **fossil record** provides evidence of change over time.

| 55–45 million years ago | 32–25 million years ago | 17–11 million years ago | 5 million years ago to present |

The second observation from the fossil record is that the earliest living things were relatively simple. Over time, living things have become more complex. For example, the oldest fossils are remains of single-celled bacteria found in rocks that are 3.8 billion years old. The first organisms with a well-defined nucleus are found in rocks that are 2 billion years old. In rocks about 540 million years old, fossils suddenly become much more abundant. Some paleontologists think the sudden explosion of fossils occurred at this time because animals began developing hard parts, like shells and skeletons. Shells and skeletons, characteristics of more complex animals, were more likely to be preserved in the fossil record than the soft parts of organisms.

How Fossils Form
Usually, when a plant or animal dies, it is completely destroyed. It may be eaten by another animal, or it may slowly decay. Sometimes the remains of a living thing are covered by **sediment**, small pieces of rock and living organisms, before they are destroyed. If conditions are just right, the animal or plant remains get preserved as fossils.

fossil record: all the fossils ever found.

sediment: small, solid pieces of material that come from rocks or living organisms.

sedimentary rocks: rocks formed from the compression and cementing together of layers of sediment deposited in oceans, lakes, and swamps.

dating: determining the age of sedimentary rocks or fossils.

relative dating: determining which fossil is older than another by comparing the relative positions of the rock layers in which they are found.

Rock layers deposited at the same time contain similar fossils. Matching rock layers around the world helps scientists date the fossils in the rocks. For example, rock layer C in Ohio matches rock layer G in New York because they contain the same fossils. Therefore, geologists would date these rock levels as being formed during the same period of geological time.

Because most fossils are made by sediment, they are found in **sedimentary rocks** deposited in oceans, lakes, and swamps. Sediment is carried into bodies of water by rivers, wind, and runoff from the land. The sediment sinks to the bottom of the body of water. This constant shower of sediment quickly covers any remains that have also sunk to the bottom. Over time, the sediment hardens into rock, preserving the shape of the living thing it covers. Because most fossils form in water, most fossils are from animals or plants that once lived in or near water.

Dating Fossils and Rocks

The fossil record includes all of the fossils ever found. Thousands of layers of sedimentary rock deposited slowly over time and covering the remains of living organisms have been found in many different places around the world. Tracing the history of life on Earth takes some detective work. First, paleontologists have to determine the age of the rocks to find out how long ago the living organism may have lived. Scientists call this process **dating**. Once the rocks have been dated, the fossils found inside them can be ordered by when they existed.

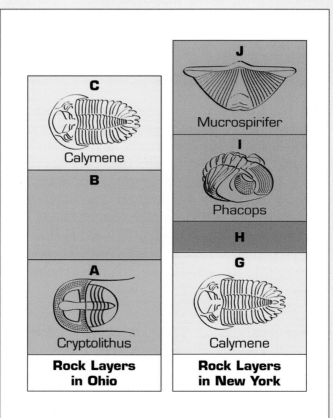

One type of dating is called **relative dating**. The geologists of Darwin's time used relative dating to determine the age of fossils. In sedimentary rock, the position of the layers tells which rocks are older and which are younger. Lower layers must be laid down before the upper layers. So fossils found in bottom layers must be older than any fossils found in the layers above.

Relative dating provides some information about the age of rocks, but more detailed and accurate information is always useful. In the 1950s, long after Darwin used relative dating, scientists introduced a new way of determining the age of rocks. They began to use a process called **radioactive dating.**

In radioactive dating, scientists determine the amount of a radioactive element in rocks and fossils. The most commonly used element is carbon 14. Carbon 14 is unstable and changes slowly into a new, unchanging element. Scientists know the time it takes for half the amount of a radioactive element to change into a new element. This length of time is called the element's **half-life**. Half-lives can be very long, more than a trillion years, or very short, much less than a second. The half-life of Carbon 14 is 5730 years. Scientists use half-life and their knowledge about how much carbon 14 is in a particular sample to calculate the actual age of a rock or fossil.

radioactive dating: a method of dating fossils by measuring the amount of a radioactive element in the fossils and in the rocks in which the fossils are found.

half-life: the time it takes for half of an amount of a radioactive element to change into a new stable element.

Plant fossils, like this leaf, may show evidence of life on Earth millions of years ago.

3.4 Investigate

How Do Adaptations and Environmental Changes Affect Traits?

You have experienced how selection pressure can change the frequency of traits in a population. When the environment changes quickly, the traits of a population of organisms can also change very quickly. In this investigation, you will simulate how the shape and structure of a bird's beak affects the type of food it is able to eat. Then you will simulate a change in the food that is available, and you will see how that results in birds with some traits dying off and birds with other traits continuing to live and reproduce.

Simulate Feeding in a Stable Environment

Predict

Your group is a population of birds of the same type that all live in a particular area. You will each be assigned a different bird beak (tweezers, toothpick, tongs, clothespin, or spoon) to use to feed. (This is similar to what Darwin observed on the Galapagos Islands. He saw finches with different types of beaks). The food choices where your population of birds lives include marshmallows, sunflower seeds, rice, and marbles. Predict which of these foods you will be able to pick up easily with your beak. Record your prediction on your *Bird Feeding Simulation* page.

Procedure

1. Tape the paper plate to the table with a loop of tape on the bottom of the plate.

2. Empty the mixture of different types of food onto the plate.

3. Take turns collecting food. When it is your turn, use your beak to collect as much food as you can in 10 seconds. You must use the beak with one hand only and must not touch the plate or food with your hand. Place the food in your cup.

4. After 10 seconds:

   - Count the number of pieces of food in your cup.

   - Record the type and number of each piece of food in the Simulate feeding section of your *Bird Feeding Simulation* page.

5. Pour the food from the cup back onto the plate. Repeat steps 3 and 4 for each group member.

6. Record each group member's data onto your *Bird Feeding Simulation* page. Have one member of your group read aloud your group data to record in a class data table.

Materials

- pair of tweezers
- toothpick
- tongs
- clothespin
- spoon
- paper plate per person
- cup per person
- plastic bag filled with a mixture of rice, sunflower seeds, marbles, and marshmallows

---

Bird Feeding Simulation 3.4.1

Name: _____ Date: _____

Simulate feeding

Prediction:

Beak Types	Number of food items eaten			
	Rice	Sunflower seeds	Marbles	Marshmallows
Tweezers				
Toothpick				
Clothespin				
Spoon				
Average (mean)				

Simulate natural selection

Prediction:

Beak Type	Number of food items eaten	
	Rice	Marbles
Tweezers		
Toothpick		
Clothespin		
Spoon		
Average (mean)		

GENETICS

Analyze Your Data

Copy the class data onto your *Bird Feeding Simulation* page. With your group, analyze the class's data and determine which beak shape was best suited for each food type. Discuss which beaks were able to pick up the greatest variety of foods. Which beak shape was able to eat the most of each food (rice, seeds, marbles, marshmallows). Which beak types were not well suited to eating particular foods?

Simulate Natural Selection in a Drought

Predict

This time imagine that the birds live on two distant islands, Island A and Island B. Each island has one bird species with each beak type. There is a drought on the islands and most of the food has died. All that is left for the birds to eat on Island A is rice. All that is left to eat on Island B is marbles. Using your data, predict which beak shape will be best at eating rice and which beak shape will be best at eating marbles. Predict what will happen over time to the birds on each island with each of the beak shapes.

You will be assigned to Island A or Island B. Use the same beak type you used before.

Procedure

1. Tape the paper plate to the table with a loop of tape on the bottom of the plate.

2. Put your type of food onto the plate.

3. Take turns collecting food. When it is your turn, use your beak to collect as much food as you can in 10 seconds. You must use the beak with one hand only and must not touch the plate or food with your hand. Place the food in your cup.

4. After 10 seconds:
   • Count the number of pieces of food in your cup.
   • Record the number of pieces of food in the Simulate natural selection section of your *Bird Feeding Simulation* page.

5. Pour the food from the cup back onto the plate. Repeat steps 3 and 4 for each group member.

6. Record each group member's data on your Bird Feeding Simulation page. Have one member of your population group read aloud your group data to record in a class data table.

Analyze Your Data

With your group, analyze the class data and consider which beak shape best allowed birds to survive on each island, and then answer the following questions:

- What do you think happened to the birds that could not easily eat the food on their island?

- What were the selection pressures on the bird species? How did the population adapt to those selection pressures?

- How did your activity simulate natural selection?

Development of Darwin's Theory of Natural Selection

On the Galapagos Islands, Darwin observed finches, a type of bird. Though the finches had many traits that were similar, many traits were different. One of the finch traits Darwin investigated was beak shape. This is a short description of Darwin's ideas about finches.

Observations: Each species of finch has a different beak shape. The mainland of South America has one finch species. The Galapagos Islands have 13 different finch species.

Question: Why does the beak shape in finches vary so much?

Hypothesis: The 13 species of finches came from the one species on the mainland. The different beak shapes are due to the different foods the birds have available on each island.

Test: Compare the traits of the Galapagos Island finches with the traits of finches on the mainland. Observe what foods each of the different finches eats.

Results: The Galapagos Island finches are different from the mainland finches mainly in beak shape. Finches on each island eat different foods, and the shapes of their beaks seem to be adapted to the type of food they eat.

Publication: "My research indicated that the 13 species of finches on the Galapagos Islands descended from the one mainland species. They became different species with different beak shapes because, over time, they competed with one another for food. Those birds with a trait for beak shape that allowed them to eat the particular foods on their islands survived. Over a long time, the population adapted to this selection for beak shape."

Other Scientists' Results: Other scientists have investigated other species. They found many similar cases of species branching off from a single species. Observations suggest that single species develop into many species through geographic separation and selection for a trait.

Theory: New species come from other species. If a population becomes divided, different parts of the population may develop into different species. They become different from one another because of different selection pressures for each part of the population. There is variation among the traits of individuals in a population, and when individuals compete for the needs of life, some individuals have traits that allow them to better compete. The fittest individuals in each population survive and reproduce. Over time, each population adapts to selection pressure through this struggle for survival.

Reflect

Discuss the following questions in class.

1. When the original finch species came to the Galapagos, it lived on each of the islands where finches are found today. Now the finches are different on each island. Draw a picture to help describe what you think happened, over time, to each type of finch.

2. One of the islands in the Galapagos has a lot of large seeds with hard coats. What do your experience and the data tell you about what will happen to a finch that eats only large seeds with hard coats if those seeds disappeared?

3. How do you think natural selection played a role in how the finches in the Galapagos changed? Describe the changes using ideas about natural selection.

What's the Point?

The traits in a population vary. Some traits allow organisms to survive and reproduce more effectively than others. Selection pressure from the environment determines which traits will allow an organism to survive. Individuals that survive can pass that trait on to their offspring. Then more individuals in the next generation will have the traits needed for better survival. In this way, and over a very long time, the population adapts. When an environment changes, the selection pressures change. Traits that once gave species an advantage can then become a disadvantage. Populations that can adapt will survive. Those that cannot adapt will die out.

3.5 Explore

How Does the Environment Affect the Growth of Rice with Different Traits?

Through the process of natural selection, organisms that are better adapted to their environment survive and reproduce. Natural selection explains the different beak shapes of the Galapagos Island finches and the variety of traits seen in many other organisms. For each population of organisms, the environment helps determine how the population will adapt. You have seen several examples of how the traits of organisms can be changed by the selection pressures of the environment.

Natural selection and selection pressures also affect the growth of rice plants. The new rice plant you have been working to develop has traits that make it able to grow in some specific environmental conditions but not in others.

Fields are flooded with water to grow rice.

One environmental factor is the amount of rainfall. Drought conditions often determine the type of rice farmers need to plant. If they know there will be drought, they plant drought-resistant rice plants. But they cannot always predict the amount of rain they will get. When they cannot predict how much rain they will get, deciding what rice to plant is complicated. Should they plant regular rice, or should they plant drought-resistant rice?

Now that you know how selection pressures affect the growth and survival of organisms, the farmers want you to advise them about which type of rice they should plant when they cannot predict the rainfall. Read the letter from the Philippine Rice Farmers Cooperative on the next page. You will be designing an experiment to help them make their decision.

Sometimes the environment changes and the fields suffer from drought.

**The Philippine Rice Farmers Cooperative,
Quezon Province, Philippines**

To: All Student Researchers

From: The Philippine Rice Farmers Cooperative,
             Quezon Province, Philippines

Subject: Request for Assistance

We have a variety of rice we want to test in the field and we need your assistance.
Most rice requires a lot of water to grow and produce seeds. Our weather is not
always rainy and sometimes does not provide enough water for the rice. These
periods of drought are very difficult for the farmers.

Some rice varieties have traits that allow them to grow during droughts and still
produce seeds. We call this type of rice drought resistant. Drought-resistant rice
plants can produce rice grains even if conditions become too dry. While these
plants still produce rice, they yield less rice than in rainy conditions.

Because the environment changes, drought resistance may be an important trait
to consider in developing a better rice plant. We would like to know if we should
plant all drought-resistant rice in our fields or a mixture of drought-resistant and
regular rice.

Please send us your recommendations on how to carry out an experiment to test
drought-resistant rice and regular rice in our fields.

Observe

Sometimes pictures make ideas more clear. You might need to know more
about how rice grows before designing your experiment. You will watch a
video showing growth of two different types of rice, one normal and one
drought resistant. Both are growing in the same field in a year when there is
about the normal amount of rain. The drought-resistant rice is growing on
the left side of the screen, and the normal rice is growing on the right side.
As you watch, notice the differences in how the two rice fields are growing.
Pay attention to how fast the rice grows on each side of the field.

Stop and Think

1. You saw two different types of rice growing in the fields. One was resistant to drought, the other was not. Compare the yields of the two fields. Which field produced more rice?

2. Which of these two types of rice would you grow if you were a rice farmer and you expected the usual amount of rainfall?

Predict

You will be designing a field experiment to answer the rice farmers' question, *How does the amount of rain affect the growth of regular and drought-resistant rice?*

Before planning your experiment to see how the environment affects the growth of rice, discuss what you think would happen if rice farmers planted all drought-resistant rice.

- What will happen to drought-resistant rice if the environment changes?

- What will happen if all the rice were drought resistant and there was a lot of rain?

- Which is more important—rice plants that produce more rice or plants that can produce rice during droughts?

Design a Field Experiment

Using a *Rice Field Experiment Planning* page, you will design experiments to investigate the effects of rainfall on drought-resistant rice and regular rice. You will send your designs to the rice farmers and they will carry out your experiments. Design one experiment to test the effects of rainfall on drought-resistant rice and another to test the effects of rainfall on regular rice. For each experiment, be certain to include the following:

Question:

What question are you investigating and answering with this experiment?

Prediction:

What do you think the outcome of your experiment will be? What could happen if the weather changes?

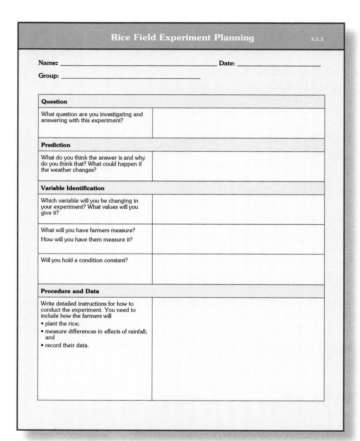

Variable Identification:

- Which variable are you going to vary in each experiment (the *independent*, or *manipulated*, variable)?

- What values are you going to give it?

- What variable are you going to observe and measure (the *dependent*, or *responding*, variables)?

- What experiment procedures and conditions do you need to control (keep the same)?

- How will you measure your dependent (responding) variables? Think about what you saw in the video to answer this.

Procedures and Data

The farmers will conduct the experiments you send them. To do this, they will need accurate and detailed procedures. Carefully develop a procedure for each of your experiments. Be sure to provide the farmers with enough information to carry them out.

The data you collect will need to be accurate and trustworthy. Be sure you can describe how your procedures will result in accurate data.

Include the following in your procedure:

- How the farmers will plant the rice so that they can compare the effects of rainfall on each.

- How the farmers will measure the differences in the effects of rainfall on the rice. Think about the video as you answer this.

- How the farmers should record the data.

You will be presenting your procedure to the class in a *Plan Briefing*. Make sure you can describe to the class why they can trust your design. Use the hints on the *Rice Field Experiment Planning* page as a guide. You will need to show the class that you have thought about all the parts of your plan.

Communicate

Plan Briefing

Two groups will now present their planned experiments to the class in a *Plan Briefing*. You will probably see the differences and similarities between these plans. In the class discussion, compare the plans to one another and to yours. Notice similarities and differences. Identify the strengths of each plan. Think about what might be improved in each.

As each plan is presented, listen for answers to the following questions:

- What is their prediction? Why did they make that prediction?

- What are the variables?

- What are the steps of the plan? Why? Are there any problems you foresee with this plan?

- What issues or concerns do they have about their plan?

- Compare their plan to yours. Are there steps they forgot? Are there steps that need to be made more specific? If so, suggest ways to make the plans better.

Revise Your Plan

Your class will send only one procedure to the farmers. Based on the discussion your class just had about the two plans that were presented, design a procedure that the class agrees on that can be sent to the farmers. The farmers will run the experiments in the fields and will send you the results to analyze later.

What's the Point?

Some genetic traits may help organisms survive and reproduce better in some environments than other traits. Organisms with traits that are better for survival have an advantage in that environment.

When considering the traits a rice plant needs to survive and produce more rice, farmers and scientists have to think about the amount of rainfall and how much rainfall varies from year to year.

Farmers have to plant drought-resistant rice plants if drought is expected. Drought resistance is a genetic trait. Since drought is always a possibility in areas where rice is grown, planting drought-resistant rice plants might be an advantage. But nobody knows how much rice drought-resistant plants produce when there is plenty of rain. Possibly the yield of the drought-resistant rice is not good when there is a lot of rain.

3.6 Analyze Your Data

Analyze Your Data From the Field Experiment

The rice farmers have conducted your experiments in the field and have sent back the results. You will now analyze their results to identify the effects of the environment on regular rice and on drought-resistant rice.

 **The Philippine Rice Farmers Cooperative, Quezon Province, Philippines**

To: All Student Researchers

From: The Philippine Rice Farmers Cooperative,
          Quezon Province, Philippines

Subject: Results of the field experiment

The results of your suggested experiment are back. Following your instructions, we planted both types of rice, each in different plots. The two plots, Plot A and Plot B, had different environmental conditions. Plot A received a lot of rain. Plot B was drier; it received very little rain.

We collected the rice from many plants of each type. We measured a sample of grains for each plant, and we are reporting two types of data: the average number of grains produced per plant per plot, and the yield of rice in tons per **hectare** from each plot. We have summarized the data in two tables.

hectare: a metric unit of area measurement equal to 2.471 acres.

**Production Data for Drought-resistant and Regular Rice
in Both Dry and Wet Environments**

Average number of grains per plant		
	Plot A Abundant rains. The fields flooded for several days during the growing season.	**Plot B** Very little rain. The fields were mostly dry for several weeks during the growing season.
Drought-resistant rice	90	120
Regular rice	120	100

Tons of rice produced per hectare		
	Plot A Abundant rains. The fields flooded for several days during the growing season.	**Plot B** Very little rain. The fields were mostly dry for several weeks during the growing season.
Drought-resistant rice	9	10
Regular rice	12	8

Analyze Your Data

The farmers grew the two types of rice in two plots with different environmental conditions. Analyze the data using the table as a guide. As you analyze the data, answer the following questions:

1. Compare the average number of grains per plant in each plot. Which rice plant produced more grains per plant in the dry environment? Which rice plant produced more grains per plant in the wet environment? Use data from the experiment to support your answers.

2. Compare the tons of rice produced per hectare in each plot. Which rice plant produced more tons of rice per hectare in the dry environment? Which rice plant produced more tons of rice per hectare in the wet environment? Use the data from the experiment to support your answers.

3. If you knew there would be a lot of rain, which rice would you plant? Why?

4. If you knew there would be drought, which rice would you plant? Why?

5. What do you think would happen if farmers always planted only regular rice in all environments? What do you think would happen if they always planted drought-resistant rice?

6. If you could not predict the rainfall, which rice would you plant? Why?

Explain

You have carefully considered the effects of rainfall on rice production. Use a *Create Your Explanation* page to help you develop an explanation that takes into account your claim, evidence, and science knowledge. Begin with your claim about the impact of the environment on rice production. The results of the field experiment are your evidence.

You also have science knowledge from the readings. To support your claim, record all this information in the appropriate boxes. Then write an explanation that uses your evidence and science knowledge to support your claim. Your explanation should include what you know about selection pressure. It should help others understand why your claim is valid.

Communicate

Share Your Explanation

Share your explanation with the class. As you listen to the explanations of others, decide if you agree with the claim that they are making. Think about how well the evidence you have seen and your science knowledge support their claim. If you do not agree or if you think the evidence and science knowledge presented do not support the claim well enough, offer advice about how to make them better. Listen carefully to the explanation given. It should connect the claim to the evidence and science knowledge that support it and convince you that the claim is valid. If you think the explanation is not accurate enough or if you do not agree with it, ask questions or offer advice. Make sure to contribute respectfully.

Monocultures

In their natural state, land areas have many plants and animals that depend on each other for survival. For example, before people from Europe came to North America and moved west, the western lands consisted of vast **prairies**. These prairies had hundreds of different species of grass and wildflowers. Each prairie plant species was adapted to the weather, the soil conditions, the animals, and the other plants in the environment. Some plants had roots that held the soil in place when the winds blew. Others provided shade from the summer Sun. The different kinds of plants depended on each other to survive. In **diverse environments** like this, the environment is balanced. Each plant and animal has a role to play.

You already know that when changes happen in an environment, the new selection pressures cause the plants and animals adapt. This can happen very quickly when people come into an environment and make many changes at the same time. For example, it is common for farmers to plow land for planting and then to plant only one type of crop. When people replace the many plant species in an environment with only one type of plant, the result is a **monoculture**. Monoculture means using the land for only one type of crop. When a monoculture is planted in place of a balanced environment, there is no longer diversity, the environment is no longer in balance, and it is hard to predict what will happen as a result.

prairie: large area of grasslands usually located in the interior of continents.

diverse environment: having many different types of organisms that depend on one another for survival.

monoculture: using the land for the growing of one crop.

GENETICS

The prairie was a rich environment that had adapted to the environmental conditions.

The period of huge dust storms in the 1930s, called the Dust Bowl, was caused by severe droughts and damage done by planting corn in the dry prairie soil.

An important example of farmers planting a monoculture happened in the western United States in the 1800s. When European farmers moved west and were looking for land to plant corn, the prairie looked like a dream to them. It had gentle rolling hills, rich soil, and abundant wildlife. They thought corn would also grow well there. So they plowed up the **native** prairie plants and planted corn. The farmers didn't know that the prairie was not a good environment for growing corn. The prairie plants had been there for thousands of years and had adapted to the soil and dry conditions. But corn was not adapted to prairie conditions. Soon the corn died from lack of water and too few nutrients.

Farming changed the prairie soil. The original prairie plants had roots holding the soil in place which made soil **erosion** much less likely. In the 1930s, farming, along with many severe droughts, caused prairie soil to blow away in great storms of dust. During this period, known as the Dust Bowl, much of the soil in these areas blew away.

After many years of failure, people have learned to grow crops on the former prairies, but it has been very difficult. In place of the prairie grasses and wildflowers are miles and miles of corn fields. Growing a monoculture is hard work. To fight the dry environmental conditions, today's farmers provide water to the crops almost every day. Farmers also have to provide nutrients and fight pests by using lots of fertilizers and pesticides. Farming the prairie has become very expensive and has made food prices higher, too.

Large areas of the western United States were affected by the storms and destruction caused by the Dust Bowl in the 1930s. People in those states were forced to abandon their homes and farms and often headed farther west.

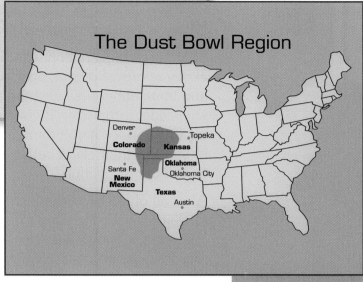

The Dust Bowl Region

Reflect

Consider what you now know about monocultures, like the corn fields planted by the farmers. Think about how the lessons the farmers learned might be applied to the rice fields.

1. How does a monoculture differ from a balanced environment? When the environment changes, what is one advantage a balanced environment has that a monoculture does not have?

2. The amount of rainfall in an environment can change at any time. What might be an advantage of planting several kinds of rice with different traits? What are some other traits the rice farmers might want their rice plants to have in addition to drought resistance?

3. What type of plants would you suggest planting in rice fields? Support your answer with evidence from the field experiment and science knowledge about selection pressure, diversity, and monocultures.

native [plants]: from the local area and best adapted to the local climate.

erosion: the loss of soil through the actions of water and wind.

Revise Your Explanation

You have read one example of the effect of monocultures on the environment and the difficulties of growing a single crop in a place that might not be ideal for it. Think about what might happen if the farmers grew only one type of rice. The environment, other plants and animals, as well as the rice might be affected.

Revise your claim and explanation so that it now includes the effects of rainfall on rice production and the possible effects of rice production on other factors and organisms in the environment. First, review and revise your claim, and then review and revise your supporting evidence and science knowledge. Include what you now know about monocultures. Finally, revise your explanation statement.

Communicate

Share Your Explanation

Share your explanation with the class. As always, as you listen to the explanations of others, decide if you agree with what is being presented and think about how well the evidence you have seen and your science knowledge support it. If you do not agree with a group's claim or explanation, or if you think a group could be more accurate or convincing or support their claim better, ask questions or offer advice. Make sure to contribute respectfully.

After all of the presentations, come to agreement as a class about a set of claims about the effects of rainfall on rice production, the evidence and science knowledge that support them, and explanations that provide reasons your claims are valid.

Update the *Project Board*

In the *What are we learning?* column, record what you have learned from the field experiment and your reading. Be sure to add the experimental evidence in the *What is our evidence?* column.

What's the Point?

A trait can give organisms an advantage under certain environmental conditions. Drought-resistant rice plants produce more rice in dry conditions but less rice in rainy conditions. The trait of drought resistance is an advantage in dry conditions but not in wet conditions.

In monocultures only one type of plant is grown. Plants in a monoculture have less variety than plants in a balanced environment. When a monoculture is planted in place of a balanced environment, the environment is no longer in balance. Much of the diversity of the original environment is lost. It can be difficult to grow crops if they are not adapted to the environment.

3.7 Read

How Do People Affect Traits in a Population?

Charles Darwin used his knowledge of artificial selection as one piece of evidence for his theory of natural selection. Artificial selection is when people breed animals and plants for specific purposes. Darwin observed many breeds, or types, of domestic pigeon species. All the different pigeon species were developed by people for specific purposes. For example, there were racing pigeons that could fly fast, carrier pigeons that could carry messages for people over long distances and return home, and pigeons that were bred because they were pretty. People had selected the traits they liked or needed and bred the birds for those traits.

How Have People Developed Different Breeds of Dogs?

Think about how dog breeding began. Wolves and early people lived side-by-side for any thousands of years, but at first they lived separately. About 10,000 years ago, wolves began to live with people. Both species benefited from living and hunting together. People provided food, shelter, and companionship for the wolf. The wolf provided hunting skills and protection for people.

Some of the traits of the wolf were useful to people, some were not. The first people to live with wolves would have kept and bred only the wolves that had desirable traits, such as good eyesight, a strong sense of smell, loyalty, protectiveness, or even swimming ability. People selected

Wolves had many desirable traits that made them useful to people. People provided shelter and food for wolves, and the wolves could hunt for animals and protect people. The traits people selected in wolves included good eyesight, a strong sense of smell, and protectiveness.

Louis Dobermann wanted a giant terrier for quickness, strength, and guard-dog qualities. The Doberman pinscher is a cross of German shepherds with German pinschers, Rottweilers, black-and-tan terriers, and greyhounds.

for traits that would match their needs and their environment. Each generation of wolves became more tame and more useful to people.

As people spread out across Earth and formed towns and communities, their needs changed. People who kept cows or sheep used artificial selection to breed animals that would protect and herd the livestock. Those who hunted for food selected for animals that would track and bring back prey. People began to select for the specific traits needed for these jobs. Through artificial selection, wolves slowly evolved into dogs, and many different breeds were developed.

Although people first bred dogs to carry out necessary tasks, today most dogs are pets, not workers. People often breed dogs for their looks and personalities. The American Kennel Club, the group that decides which dog breeds are true breeds, recognizes over 150 breeds of dogs.

Producing a new dog breed takes many years and generations. The artificial selection of dogs can be accomplished in many ways. First, dogs are selected for desirable traits like body type, color, speed, herding or hunting ability, endurance, and size. For example, the greyhound was selected for a slim body, long legs, and speed. Sometimes breeders look for puppies with an unusual trait, such as hairlessness or lack of a tail. These traits generally don't help dogs survive in the wild, but they can give a dog a unique look that appeals to breeders and pet owners.

New breeds can also be made by cross-breeding two different true breeds. The offspring will not look or behave like either parent.

Reflect

1. What traits would you want a dog to have if you planned to use the dog as a guide dog for a blind person? What traits would you not want it to have? How would you select for the desirable traits?

2. How might breeding a dog for specific traits be similar to breeding rice for specific traits? What factors would you need to think about? Using the steps people use to produce new dog breeds, discuss the steps you would take to breed a new rice plant with specific traits.

How Have People Developed Variations in Plants?

maize: another name for domesticated corn.

teosinte: the wild ancestor of corn.

People also use artificial selection to breed plants for food. People have been breeding plants for as long as they have been breeding animals. One important plant people have been breeding is corn. People began to grow corn in a monoculture about 200 years ago.

The kind of corn you are familiar with is called **maize**. Maize was developed from a North American plant, **teosinte**, which grows in the wild in Mexico. There is evidence that Native Americans had begun growing this wild plant in their gardens and fields over 1800 years ago.

Over time, people have changed teosinte in several ways. Maize yields more food than teosinte because the grains on each ear are bigger and there are more rows per ear. The casing around each grain is softer, making it easier to eat. The husk protects maize in bad weather.

Teosinte, an ancestor of modern corn, grew wild in Mexico and was domesticated by Native Americans.

However, the breeding of traits for food has made maize more difficult to grow. Maize is not resistant to pests and diseases, and it requires more water and fertilizer than teosinte. Scientists are working to cross the traits of teosinte and of maize to create a plant that resists disease and pests but provides good nutritional value. In 1977, a Mexican biologist discovered a wild teosinte species with genes that protect it against many diseases that affect maize. By breeding that teosinte species with maize through artificial selection, scientists have developed some disease-resistant maize varieties.

Scientists are still working on developing varieties that are easier to grow and provide good nutrition.

GENETICS

Modern corn is the familiar name for the grain also known as maize.

Reflect

1. List several traits humans might select for when breeding different varieties of rice.

2. How might breeding different varieties of rice through artificial selection benefit the farmers in the Philippines?

Update the *Project Board*

Record new science knowledge from your reading about artificial selection in the *What are we learning?* column. Record evidence from your reading to support your new knowledge in the *What is our evidence?* column. Record in the *What do we think we know?* column what you think you know about how people can act as selection pressure on a population. You may also have come up with other questions or ideas for investigations. Record these questions and ideas in the *What do we need to investigate?* column.

What's the Point?

People can affect traits in a population of plants or animals. Many animals and plants we know today became different breeds or varieties through artificial selection. Each breed or variety came from a single wild ancestor. People selected for traits they wanted and bred individuals for those traits. This selection took many years and many generations. Scientists today are still investigating to find out how they can use selection to improve domestic species of plants and animals.

3.8 Explore

How Can Artificial Selection Produce Individuals with Specific Traits?

In natural selection, the selection pressure comes from the environment. In artificial selection, the selection pressure comes from people. Both these processes result in changes in the traits of a population. Natural selection selects traits that allow organisms to survive in their environment. Artificial selection changes a population to fit criteria set by people. These criteria may have nothing to do with the environment. In fact, most domestic species, bred through artificial selection, could not survive and reproduce very well without people to care for them.

A dog breeder's goal is to develop a breed of dog that meets his or her criteria. One generation at a time, dogs are bred closer to the desired traits. You are now going to work as a bird breeder trying to produce a bird that meets specific criteria. You will simulate breeding birds, repeating the type of genetics experiments Mendel did when he crossed pea plants.

In this simulation, breeding the birds to match your criteria happens very quickly. In real life, breeding birds with specific traits does not happen as quickly. While you work at this simulation, imagine what it would be like to have to breed birds over many generations. Think about the costs and the patience needed to select just the right male and female birds to breed.

Guide dogs for blind people are bred for very specific criteria. The dogs must be highly intelligent, calm, patient, and loyal. Artificial selection by breeders helps to develop individual dogs with the desired traits.

Your Population Model

The fantasy birds you will breed have many different traits. Some traits are more desirable than others to the people who want to buy your birds. You need to select and breed only birds with the most desirable traits.

Your fantasy birds have these genetic traits:

Bird Breeding Information Table			
Trait	**Genotypes and Phenotypes**		**Most desirable trait**
Crest color (Only males have crests, though females may carry the allele for crest color.)	Gray AA or Aa	Blue aa	Blue
Wing Color	Gray BB or Bb	Red bb	Red
Tail Color	Gray DD or Dd	Red dd	Red
Breast Color	Gray CC or Cc	Purple cc	Purple
Sex	Male XY	Female XX	Male
Desirable Bird	Male with a blue crest, red wings, red tail, and purple breast		

Like all breeders, you have a goal. You want your bird to have all the most desirable traits. You want to breed a male bird with a blue crest, red wings, red tail, and a purple breast. Once you get started with the simulation, the most desirable traits are listed at the bottom left on the simulation window.

Most desirable bird.

Birds can have four different traits.

Project-Based Inquiry Science

Before you can create your most desirable bird, you need to know which of these traits are dominant and which are recessive. Recall from earlier investigations that the dominant alleles of a gene are represented by an upper-case letter, such as *D*, while recessive alleles are represented by a lower-case letter, such as *d*. Each gene, or trait, has two alleles, in this case, *DD*, *Dd*, or *dd*. The dominant trait is expressed if the gene has one upper-case letter (*DD* or *Dd*). The recessive trait is expressed only if the gene has two lower-case letters (*dd*).

Plan

In this simulation, you will use artificial selection to produce three birds with all the desired traits, and then you will sell those birds. Use the *Bird Breeding Information Table*, and begin to think about which birds you will need to breed to obtain the desirable traits. Begin by identifying the desirable traits that are dominant and those that are recessive. Predict which crossings will produce each of the desirable traits. This is very important in your breeding. Pay attention to the ways you might be able to get a bird with a blue crest. Record your predictions on a piece of paper so you can come back to them later.

Get Started

You will be using the computer program, NetLogo, to run your simulation. The first step is to set up your model in NetLogo. Follow the instructions in the box below to do that. You will begin with $500 to spend on breeding. It will cost you $10 each time you breed a pair of birds. You will be able to purchase birds from bird breeders for $10 each. After you set up the model, you will explore how the simulation works.

Your goal is to breed 3 desirable birds.

Be a Scientist

Using NetLogo to Set Up Your Model

Once again you will be using the computer program, NetLogo, to investigate how genetic traits in a population can be affected by selection pressure. This time, as a bird breeder, you will provide the selection pressure. Open the program on your computer. Your teacher will tell you which model to open.

Once the model is loaded, you should see a screen like the one shown. The screen has four main sections:

- The *Model Setup* settings are on the center left of the screen.
- The *Breeding Options* buttons are at the bottom left of the screen.

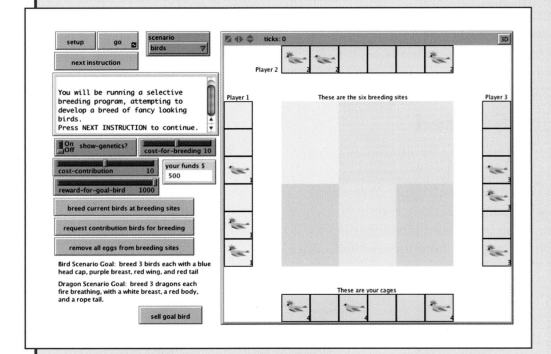

- The *Model Field* is to the right.
- The *Control Buttons* are on the top left of the screen.

The *Model Setup* settings are where you set the parameters for your simulation.

- First, click the *setup* button at the top left of the screen.

Project-Based Inquiry Science

- *Scenario*: This button, next to the *setup* button, shows which simulation you will be running. This button should be set to birds.

- *Show-genetics*: This slider button allows you to show the exact genotypes of the birds. Slide this button up for ON. Knowing the genotypes of the birds will help you decide which birds to breed.

- *Cost-for-breeding*: This slider button sets the cost for every pair of birds you breed. This value should be set at $10.

- *Cost-contribution*: This slider button sets the cost for every bird you buy from other breeders. This value should be set at $10.

- *Reward-for-goal-bird*: This slider button sets the amount you make for selling three birds with the most desirable traits. This value should be set at $1000.

- *Your funds*: This window shows the amount of money you have to breed your birds. This value should be set at $500.

Look at the *Breeding Options* buttons. You will use them to make decisions about breeding your birds, buying additional birds, and selling your birds.

- *Breed current birds at breeding sites:* This button allows you to breed your birds after you select the birds to breed. Every time you click this button, you will be charged $10 for each pair of birds that breeds.

- *Request contribution birds for breeding:* This button allows you to buy additional birds from other breeders. When you click this button, you will receive birds at random. You cannot choose the birds you want. For every request, you will receive three new birds. Every time you click this button, you will be charged $10 for each new bird, or $30 total.

- *Remove all eggs from breeding sites:* If you want to remove all the eggs you have in your breeding sites, click this button.

- *Sell goal bird:* Click this button when you want to sell your birds. You can sell one bird at a time or wait until you have three.

View the *Model Field*. The *Model Field* is where you keep your birds. This field contains three areas.

Birds you own: The six cages at the bottom of the field contain the birds you own. You are breeder number 4. This number is shown in the bottom right corner of each cage with a bird in it. You will start the simulation with three birds. Each cage cannot hold more than one bird at any one time.

Birds other breeders own: The cages around the perimeter are the cages holding the birds other breeders own. You cannot use these birds for breeding unless you purchase them. The other breeders are numbered 1 to 3. These numbers are shown at the bottom corner of each cage with a bird in it.

Breeding locations: The middle of the field contains six breeding cages where you will put birds for breeding. Each breeding location has only six spaces. These spaces can hold birds or eggs.

The *Control* buttons begin and end the simulation.

setup: Click this button to set your beginning parameters and to reset them after running a simulation.

go: This button begins your breeding simulation. Click this button after you have set your parameters. Click it again to end your simulation.

Next Instruction: This window shows you the next step in the simulation and may help you carry out the simulation. You can use it at any time.

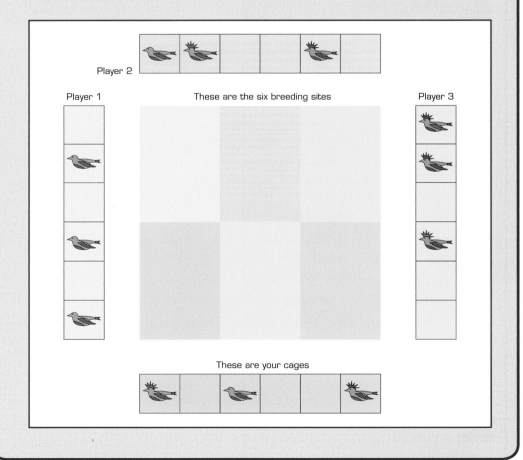

Explore the Simulation:
Breed a Bird with a Red Tail

To find out how the simulation works, you will use the simulation to breed a bird with a red tail. Think about different ways to breed a bird with a red tail. Keep these points in mind as you observe how the simulation works. As you work, identify:

- the process of breeding your birds in the simulation

- the differences between male and female birds

- how you select birds to breed for the next generation

- how you buy birds to provide more stock for your breeding

- how you move birds around

- how you release birds

- how you hatch eggs

Procedure

1. Check the *Model Setup* settings at the left center of the screen to be certain they are set correctly. These are your beginning parameters. If they are not correct, click the *setup* button and enter the correct parameters.

2. Click the *go* button to start the simulation.

3. Look at the birds in your cages at the bottom of the field. Observe which traits they show (phenotype) and their genotypes.

These are your cages

Player 4

Your birds.

If you cannot see the full genotype of each bird, you can move the bird by clicking on it, holding down the mouse button, and dragging it partially out of its cage. Do not release the mouse button while the bird is in the white space or it will fly away!

GENETICS

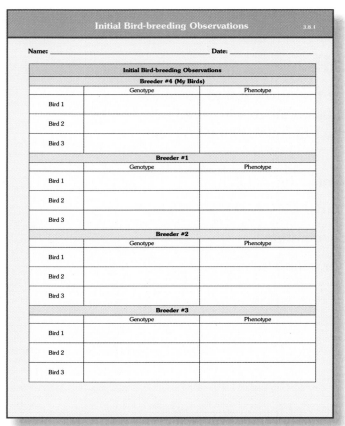

Initial Bird-breeding Observations 3.8.1

Name: _____ Date: _____

Initial Bird-breeding Observations		
Breeder #4 (My Birds)		
	Genotype	Phenotype
Bird 1		
Bird 2		
Bird 3		
Breeder #1		
	Genotype	Phenotype
Bird 1		
Bird 2		
Bird 3		
Breeder #2		
	Genotype	Phenotype
Bird 1		
Bird 2		
Bird 3		
Breeder #3		
	Genotype	Phenotype
Bird 1		
Bird 2		
Bird 3		

Bird-breeding Crossing 3.8.2

Name: _____ Date: _____

Results of Punnett square and actual crossings for red tail

Punnett square crossings that will produce a blue crest

Punnett square crossings that will produce red wings

Punnett square crossings that will produce a red tail

Punnett square crossings that will produce a purple breast

4. In your *Initial Bird-breeding Observations* page, list the genotypes (*DD*, *Dd*, or *dd*) for tail color for each bird you own. Use the *Table of Traits* to help you. If you already have one or more birds with a red tail, try to breed another one.

5. If you think you have two birds you can breed for a red tail, perform a Punnett square in the first box of your *Bird-breeding Crossing* page to cross these birds for this trait. If you do not have two birds to cross, read the section *Buying Birds from Breeders.*

6. If the results of the Punnett square were not what you expected, select two different birds and try another cross.

7. If the results of the Punnett square are what you expected, drag the male bird

you want to breed into a breeding location by clicking on the bird you want to move, continuing to hold down the mouse button, moving the mouse to the breeding location, and letting go of the mouse button. You then drag the female you want to breed into the same breeding location. Click the *breed current birds at breeding sites* button. Remember, you must breed one male bird with one female bird. If you've made a mistake and put two male or two female birds into a breeding location, release the bird that you do not want to breed by using the mouse to drag it back into one of your cages and letting go of the mouse button.

8. Your female bird will lay four eggs. Drag each egg into one of your cages at the bottom of the window. When an egg is placed in a cage, it hatches, and then you will be able to see the outcome of your cross. If there are more birds than you have room for in your cages, drag some of the birds from your cages into breeding locations to make more room.

9. Record the results of this crossing in the first box of your *Bird-breeding Crossing* page. Compare the traits of the birds that hatched with what you predicted in the Punnett square. Were you successful at breeding a bird with a red tail? If not, try the process again until you are able to breed one bird with a red tail.

10. Trying the process again will be a little trickier. You might not have enough space in your cages to save all of the birds you are not breeding. You have two options. You can drag a bird into a breeding box to save it, or you can release some birds. To release a bird, drag it into the white space, and release the mouse.

You need to be careful doing both of these things. You won't want to release a bird that you might need later, and you will want to make sure you aren't breeding birds you don't want to breed.

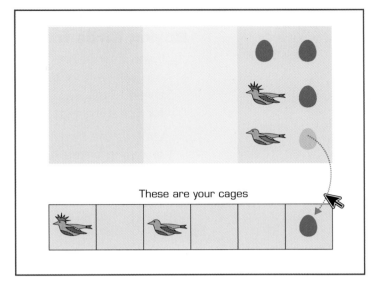

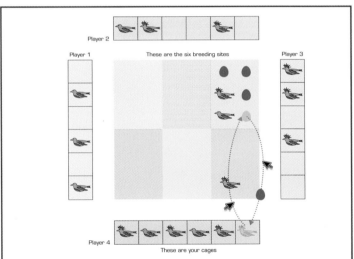

Moving a bird to make room for an egg

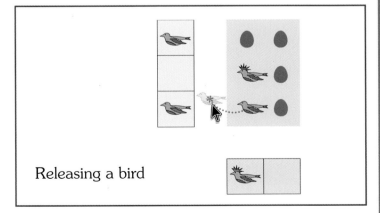

Releasing a bird

Buying Birds from Breeders

1. If you cannot produce the traits you want with the birds you have, you will have to buy more birds. Each purchase will cost you $30 for three birds. To purchase birds, click the *Request contribution birds for breeding* button.

2. Look at the birds the other breeders have sold you. They will be in breeding boxes. Observe where the new birds go and what traits they have before breeding them. Record the genotypes for tail color of those birds in your *Initial Bird-breeding Observations* page.

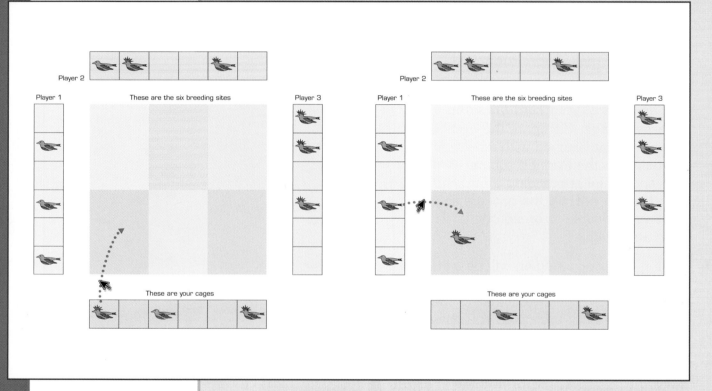

3. If you think you can cross one of your birds with a bird from another breeder and produce a bird with a red tail, go back to Step 4 of the procedure and breed your birds.

4. If you think you still cannot breed a bird with a red tail, purchase 3 more birds, and then go back to Step 4 of the procedure.

Communicate

Share with your class your experiences in breeding a bird with a red tail. Report anything you discovered that you think will make it easier for others to use the simulation. Ask questions about any parts of running the simulation or any ideas about breeding that are confusing.

Breed Three Desirable Birds

The goal is to breed three birds with the desired traits. Now that you've been able to breed one bird, you should be prepared to run the simulation and produce three male birds with the four desired traits: blue crest, red wing, purple breast, and red tail.

As before, work with one trait at a time, and be careful about how you cross your birds, which birds you select for your crossings, which birds you keep, and which birds you release.

Procedure

1. Click the *setup* button to reset your parameters.

2. Click the *go* button to begin the simulation.

3. Use a new *Initial Bird-breeding Observations* page to record the genotypes and phenotypes of all the birds in the field. Use *Bird-breeding Crossing* pages to decide which birds to cross.

4. Each time you perform a breeding, record the results in a *Bird-breeding Results* page. You may have to use several pages.

5. When you have produced three birds with all the desired traits, move them into your cages at the bottom of the field and press the *sell goal bird* button.

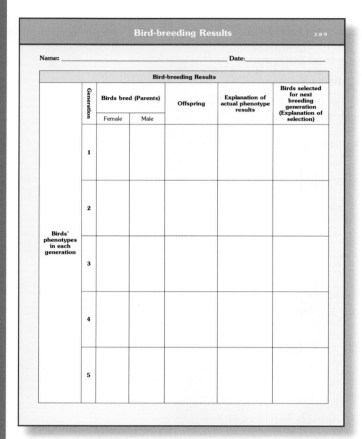

Analyze Your Data

In your group, discuss how you bred your birds. Be prepared to share your answers with the class.

- How did you select birds for breeding?

- What were the results of each of your crossings? How did your actual results differ from the Punnett squares you performed?

- How did you decide which birds to keep and which to release?

- How many generations did you breed before you produced one bird with all the desired traits? How many generations did you breed before you produced all three birds with the desired traits?

- If you ran out of money or time before you produced three birds with the desired traits, why do you think you were not successful? What could you have done differently to be successful?

Communicate Your Results

Investigation Expo

Create a poster describing your breeding. Describe how you achieved the goal of breeding three birds with the required traits. Include enough detail in your poster so your class will understand what you discovered about how to breed birds with specific traits.

One way to present your data would be to include your *Bird-breeding Crossing, Initial Bird-breeding Observations,* and your *Bird-breeding Results* pages. You may want to include colored drawings of your birds from each breeding to more clearly show how you chose which birds to breed and the results from those crossings. Your poster should include

- how you selected birds for breeding

- how selection pressure caused the traits of the population to change

- how difficult it was to breed birds for specific traits

- how knowing the genotypes of the birds helped you select birds for breeding

- how Punnett squares helped you decide which birds to breed

- how many generations you had to cross to produce one bird with the desired traits

- how many generations you had to cross to produce three birds with the desired traits

- how your actual results differed from your prediction

As you walk around the room and look at the posters of other groups, observe how they selected their birds for breeding and the results they produced. Observe how many generations of birds they crossed before they produced the desired traits. Compare their results with yours. How do they differ from yours? Ask questions if you do not understand their results or how others carried out their simulations. Be sure to ask your questions respectfully.

Reflect

1. Real bird breeders often do not know the genotypes of all the birds in the population. How do you think they select birds for breeding? If you did not know the genotypes of the birds, how would your breeding plan have to be different from what you did?

2. What made this simulation a model of artificial selection and not one of natural selection?

3. In this bird-breeding simulation, you observed that each time you selected for one trait, other traits changed. Why do you think this happened? How might this result affect rice breeding?

4. How are breeding birds for desirable traits and developing new varieties of rice similar? How are they different?

5. Rice farmers want to produce a new rice plant from two varieties that have different traits. Which would be more efficient (take less time)— artificial selection or natural selection? Why?

What's the Point?

Artificial selection is similar to natural selection in that it causes changes in a population because of selection pressure. In artificial selection, people choose individuals to cross to produce specific traits. People apply the selection pressure on the population. Since they often don't know the genotypes of the individuals they cross, people choose individuals by their traits, the expression of the genes. Each time people select for one trait, other traits change also. This is because an organism has an entire genotype, made up of many genes, not just one. It can take many generations to produce an individual with all the desired traits.

Llamas, members of the camel family, have been domesticated for over 5000 years. They are bred as pack animals, can pull carts, and guard livestock. Llamas are often bred for their wool. Their wool may be solid, spotted, or marked in a wide variety of patterns, with colors ranging from white to black and many shades of gray, beige, brown, and red.

Learning Set 3

Back to the Big Challenge

*Make Recommendations About Developing a
New Rice Plant That Will Produce More Rice
and More Nutritious Rice*

RBWI
The Rice for a Better World Institute

To: All Collaborating Scientists

From: The Rice for a Better World Institute (RBWI)

Subject: Research Update

The work you have done for the *RBWI* and the suggestions you have made for developing a new variety of rice have been very important. We are constantly concerned about improving the rice crop so the people will have enough rice to eat and that rice will be very nutritious.

Because you now know more about how genes work and how the environment affects how traits are expressed, it would be helpful if you would update your previous recommendations for the new rice plant. We again include the table of the traits to consider.

Rice variety	Trait	Inheritance
A	grows well in dry conditions	recessive
B	grows well even in flood conditions	dominant
C	has high starch content	dominant
D	has high fiber content	recessive
E	has high levels of vitamins and minerals	recessive
F	is resistant to pests	recessive
G	is resistant to disease	recessive
H	produces more rice grains per plant than other rice	recessive
I	requires less fertilizer per acre of rice than other rice	dominant

Thank you for your continued work.

GENETICS

Recommend

At the end of *Learning Set 2*, you created three recommendations for the scientists and farmers. These recommendations focused on developing a new rice plant. Now that you know much more about the genetic makeup of rice and how the environment interacts to change the plants, you will be able to update your recommendations.

Review the table of traits the *RBWI* has sent with this new information about genes. Read the table carefully and discuss with your group the outcomes of growing rice with particular traits. Think about what would happen to the amount of rice grown if the rice is more resistant to pests or disease. Think about how you would like to make more nutritious rice: more starch or more different types of vitamins and minerals.

Review the three recommendations you made at the end of *Learning Set 2*, this time considering how organisms with particular traits interact with the environment. Revise each recommendation, using what you know now about the interactions of traits with the environment. Use a new *Create Your Explanation* page if you need to rewrite a recommendation.

Your first recommendation should focus on which rice types are most important to grow. Focus on your trait (more rice or more nutritious rice) and determine the rice variety, or varieties, you think the farmers should grow. Use the table of traits to help you make your decisions. Think about the ways the environment might influence how that trait is expressed.

Farmers must make sure that rice plants have the correct nutrients to grow. They also must help protect the growing plants from pests. Often, farmers will spray both nutrients and pesticides on the rice plants.

Support your first recommendation with evidence to help the *RBWI*'s scientists and farmers know why they should trust it. Think about starting your recommendations with *If, When,* or *Because*. For example, you might begin a recommendation by writing, *"If the farmers wanted to plant seeds that would produce rice that was more nutritious, they should plant seeds that..."*

Your second recommendation focuses on crossing types to create a rice that has traits even more useful for addressing the challenge. Again, review the various rice types with your group, thinking carefully about which rice types you might be able to combine successfully and obtain even better traits. Keep in mind the interaction of the traits with the environment. As you write this recommendation consider beginning it with *Because*.

For example, *"Because resistance to pests and disease will be good for growing more rice..."* Then complete the statement. As you work on your recommendations, keep in mind the ideas of Darwin and natural selection as well as the impact of artificial selection.

Finally, re-read and consider how you can update your third recommendation. Add to your recommendation evidence from your reading in *Learning Set 3*. Begin this recommendation with *"When farmers use the recommended plants, they should..."* Remember to include the expected results of their planting.

As you update your recommendations, keep in mind these questions from *Learning Sets 2* and *3*:

- Which traits will you be selecting? Why did you select those traits? Are those traits dominant or recessive?

- How will the traits of your rice be passed to the next generation? How will you make sure your new rice contains the traits you want it to have? How will this affect your recommendations?

Create Your Explanation

Name:_____ Date:_____

Use this page to explain the lesson of your recent investigations.

Write a brief summary of the results from your investigation. You will use this summary to help you write your Explanation.

Claim – a statement of what you understand or a conclusion that you have reached from an investigation or a set of investigations.

Evidence – data collected during investigations and trends in that data.

Science knowledge – knowledge about how things work. You may have learned this through reading, talking to an expert, discussion, or other experiences.

Write your Explanation using the *Claim*, *Evidence*, and *Science knowledge*.

- How should the farmers breed their plants? What environmental factors will you need to take into account?

- What selection pressures will come from the environment? How will these selection pressures affect the new rice variety? What selection pressures will you be able to control for? How will you control for these selection pressures? What selection pressures will you not be able to control for?

- What selection pressures could come from people? Which of these selection pressures will you be able to control for? How could these selection pressures affect the new rice variety?

Communicate Your Solution

Solution Briefing

After you have developed your recommendations, you will communicate them to one another in a *Solution Briefing*. Use the following questions to plan your presentation.

- Which traits will your rice need?

- What environmental factors did you take into account? How will the traits of your rice address the environmental concerns?

- How did selection pressures from the environment and people affect your recommendations?

- How does your rice meet the criteria?

- How did constraints affect your recommendations?

- What information did you use to support your recommendations?

- What ideas did you think about along the way, and why did you not recommend them?

- What questions do you still have?

As you listen to the other presentations, make sure you understand the answers to these questions. If you do not understand something, or another group did not present something clearly enough, ask questions.

You can use the questions as a guide. When you think something can be improved, be sure to contribute your ideas. Be careful to ask your questions and make your suggestions respectfully. As you listen, record your notes on a *Solution-Briefing Notes* page.

Reflect

1. How did your recommendations differ from those of other groups who worked with the same criterion? How were they similar?

2. How would you combine all the recommendations for the criteria of the challenge to make one recommendation?

Recommend

As a class, come to an agreement on recommendations for a rice-breeding plan that combines the two traits you desire in a new rice plant: more rice and more nutritious rice.

Update Criteria and Constraints

Now that you've learned more about addressing the challenge, you may realize that the criteria and constraints are different from what you first expected. Using your new knowledge and evidence from this *Learning Set*, review your list of criteria and constraints. Update the list, making it more accurate.

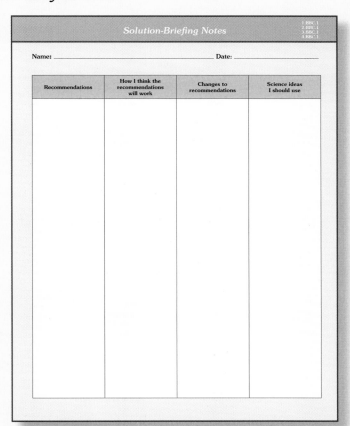

Update the *Project Board*

The last column on the *Project Board* helps you pull together everything you have learned in the Unit. The last column of the *Project Board* is the place to record what you have learned that might help you address the *Big Challenge*, to make recommendations on developing a new rice plant that will produce more rice and more nutritious rice. This column is also the place to record recommendations about how to address the challenge. Add your recommendations to the *Project Board* so you can return to it when you address the *Big Challenge* and answer the *Big Question* at the end of the Unit.

Learning Set 4

How Do Cells Grow and Reproduce?

In *Learning Set 3*, you saw several examples of how the traits of a population can change as a result of selection pressure from the environment. The traits of organisms have been evolving for millions of years. Because traits are expressed in an organism's phenotype, those changes are often visible. Both Mendel and Darwin observed the ways phenotypes changed over time. Each of them developed theories about how traits change. But neither Mendel nor Darwin had the tools to see what happens inside cells. Therefore, neither of them knew what processes cause phenotypes to evolve.

Scientists now know that information for traits is located in alleles that make up the genes on an organism's chromosomes. The environment selects for particular alleles under different environmental conditions, and the most desirable traits are passed from parents to offspring.

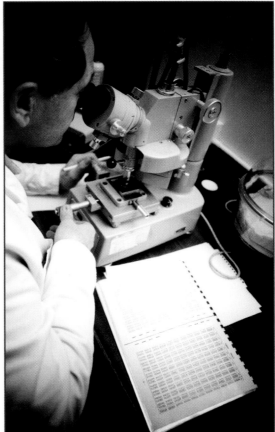

Scientists have tools that allow them to understand alleles, genes, and chromosomes. They can even see how chromosomes behave as new cells and organisms are being produced. To get the whole picture of how traits in populations change over time, you need to look deep into the cell. By looking inside the cell, scientists have explored changes in traits.

Begin by reading the announcement from the research institute to find out the next steps for your work. The scientists at *RBWI* want you to learn about the processes happening in cells. Then you can tell them how to choose seeds that will grow into rice plants that are resistant to caterpillars.

RBWI
The Rice for a Better World Institute

Research Announcement

To: All Collaborating Scientists

From: The Rice for a Better World Institute (RBWI)

Subject: Research Update

Thank you for your recommendations about how to breed white rice from red rice. The farmers carried out your breeding plan, and they were able to grow more rice and more nutritious rice. One rice variety even achieved both criteria. This result is very promising. However, in some of our fields, caterpillars still ate the rice plants. We do not think we can do any further breeding to make the rice plants resistant to the caterpillars. We hope, however, that if we know more about what is happening in the cells of the plant, we will be able to identify rice seeds that will produce plants resistant to the caterpillars. Please explore plant cells and the processes that transfer traits in the genes. After you learn more about those processes, we will ask you to help us make the rice plants resistant to caterpillars.

4.1 Understand the Question

Think About how Organisms Grow and Reproduce

There are many reasons why cells need to reproduce (make new cells). Every day, some cells in an organism die. Cells can die from illness, from damage, or from old age. Cells need to be replaced when they die. New cells need to be duplicates of the cells that died. In a person, when a skin cell dies, the person needs another skin cell, and when a muscle cell dies, the person needs another muscle cell.

Another reason living organisms need to make new cells is to grow. Organisms begin small, usually as one cell. They need to produce new cells to grow. For example, as you grow, your muscles and bones need to grow larger and stronger. For that, your body needs to make more cells.

It is the same in other living organisms, both plants and animals. In plants, when cells die, they are replaced by new cells. Cells that make and store the plant's food are replaced by other cells that can make and store food. Water-conducting cells are replaced by other cells that allow water to flow to different parts of the plant. Plants also need to make new cells to grow larger and mature. A plant begins as one cell. It needs to produce cells to develop its parts and so that each of those parts can grow larger and mature.

You also know one more reason cells reproduce—to make new organisms.

The *RBWI* scientists want you to learn about the processes going on in cells that determine what traits organisms have. You know some things about how new organisms get the traits they have. You know that male and female sex cells each carry alleles, and that when they merge to make a new organism, the combinations of alleles determine what traits the new organism will have. However, you don't know yet about what actually happens inside cells for this to happen. In order to advise the *RBWI* scientists about how to identify rice seeds that will produce plants resistant to caterpillars, you will need to look inside cells and observe the processes that allow them to grow and reproduce.

Observe

Video of a Living Cell

In this investigation, you will watch a video of a living cell as it makes a new cell. This video was made using a microscope with an attached video camera. You cannot see the details of a cell with your eyes alone.

As you watch the video, notice the chromosomes in the center of the cell. The chromosomes have been dyed to make them visible. The dyed chromosomes appear as strands inside the cell. You will watch the entire video several times to identify the chromosomes and see what happens as the cell divides.

Procedure

1. Watch the entire video. Identify the chromosomes.

2. You will watch the video a second time and draw what you see. But first, read the directions below. They will help you know what details you should be looking for in the video.

   • Look at how the chromosomes are arranged at the start of the video. How are they arranged around the cell? Draw a circle on paper. The circle represents the cell. Inside this circle, draw the chromosomes as they appear at the start of the video. Label your sketch *Chromosomes at the start of cell division*.

   • Observe the position of the chromosomes at the end of the video. How did the position of the chromosomes change? Draw another circle on a second page. Draw the chromosomes as they appeared at the end of the video. Label this drawing *Chromosomes at the end of cell division*.

   • Identify the different arrangements of the chromosomes shown in the video. Make additional drawings showing the arrangements of the chromosomes as they move from their position at the start of the video to their position at the end of the video. If you need to see the video again to complete your drawings, ask the teacher to show it again.

3. Watch the video again. Draw several more pictures to show the movement of the chromosomes from the beginning to the end of the process. Remember to label each drawing.

4. When you think you have identified all of the arrangements and drawn and labeled them, put them in order from beginning to end, and be prepared to share them with your class.

Conference

With your group, discuss the pictures you drew and labeled. Identify the different arrangements of chromosomes you saw in the video. Describe the arrangements of the chromosomes in each drawing. For example, chromosomes may be arranged in a circle in the center of the cell, or they may appear arranged along a line. Draw each arrangement, and record a description of each one. Then organize your drawings from start to finish.

Cell duplication is a process that occurs in a specific way. Scientists divide this process into steps to help them better understand and talk about what happens during cell duplication. In each step, the chromosomes move in specific ways that allow a new cell to have the same number of chromosomes as were in the original cell. To help you understand the process, label your drawings Step 1, Step 2, Step 3, and so on. Be prepared to share your descriptions and drawings with your class.

As you meet with your group, discuss these questions:

- Why do you think the chromosomes move from their position at the start of the video to their position at the end of the video?

- How would you describe the movement? What kind of patterns did you see? What evidence do you have to reach that conclusion?

- You know that cells reproduce for three reasons: to make new cells that can replace dying cells, to make new cells so an organism can grow and mature, and to make new organisms. Which parts of those processes did you just observe? What other cell processes do you need to know about to understand how these three things happen?

GENETICS

Communicate

Share your group's arrangements with the class. Describe how you identified the different steps, and describe how the chromosomes are arranged in each drawing. Pay attention as other groups present their drawings and descriptions. Did other groups find the same steps? If not, how are their steps different from yours, and how did others describe what was happening to the chromosomes in their drawings?

After all the groups have presented, spend some time as a class discussing what you just saw. What other things do you still need to learn more about to understand how cells grow and reproduce?

Update the *Project Board*

During your class discussion, you may have discovered that there are now things you think you know about how cells make new cells. Record what you think you know in the *What do we think we know?* column of the *Project Board*. You probably also identified cell processes you don't understand well yet. You might have identified questions you need to answer to be able to tell the RBWI scientists how to develop rice plants that are resistant to caterpillars. Use the *What do we need to investigate?* column to record these questions. You will be answering questions about how cells grow and reproduce in the rest of this *Learning Set*.

What's the Point?

An organism makes new cells to replace cells that have died or are damaged. Organisms also make new cells so the organism can grow. New cells must have all the genetic material from the chromosomes to function. That means that when cells reproduce (make new cells), the genetic material must be transferred to each new cell. During the process of cell duplication, the chromosomes and other cell parts move in definite patterns.

4.2 Explore

How Do Cells Divide?

A rice plant, like a person, is made up of billions of cells. Each plant develops from a single, fertilized cell in a single seed. Each time a cell divides, all of the information in its chromosomes must be transferred to the new cells.

The video you watched showed the process of cell division. You were able to capture many of the steps of the process in your drawings, and you put those drawings in order. The action in the video happened quickly. To better understand the process, you might find it helpful to slow it down.

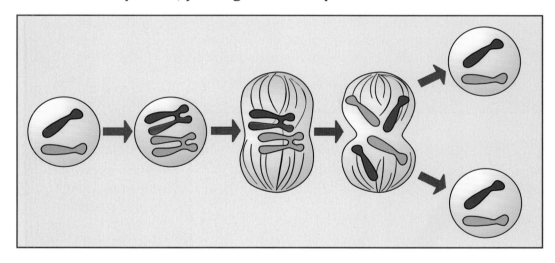

Materials
- microscope
- prepared slides of onion root tip
- pages for sketching

Observe Cell Division with a Microscope

In this investigation, you will use a microscope to carefully observe onion root-tip cells. Everyone in your group will have a chance to look at these plant cells.

In the video, you observed living cells as they were actually dividing. The slides you will be looking at show preserved cells. These are cells that have been *frozen in time* during the process of cell division. Different cells on the slides show different arrangements of chromosomes. You will be examining the slides to find examples of each of the arrangements of chromosomes in cell division. You will try to match what you see on the slides to the pictures you drew as you watched the video. Follow the procedure on the next pages as you examine your slide.

Be a Scientist

The Microscope and Magnification

You cannot see most cells with the unaided eye; you need a microscope to observe them. There are different types of microscopes. The type you will be using is called a compound-light microscope. The diagram shows the parts of this type of microscope.

The *stage* holds the microscope slide. The *clips* hold the slide in place. An opening in the stage lets light shine through. The *ocular lens* magnifies the object, usually 10x. The *objective lenses*, found on a *revolving nosepiece*, also magnify the object. There are usually three objective lenses. The shortest, or *low-power lens*, usually magnifies the object 4x. The *medium-power lens* magnifies the object 10x. The longest lens, the *high-power lens*, magnifies the object 40x. The *coarse-adjustment knob* moves the stage up and down and is used only with the low-power lens. The *fine-adjustment knob* also moves the stage up and down. It is used with the medium- and high-power lenses.

You may have used a hand lens (sometimes called a magnifying glass) to make objects appear larger. An important development of the microscope was to use two lenses. If a lens magnifies an object 10x, the object appears 10 times larger. If you then add another lens that also magnifies the object 10x, the object now appears 100 times larger (10 x 10).

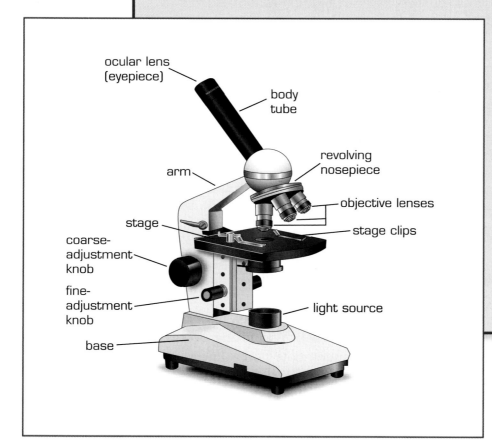

ocular lens (eyepiece)

body tube

revolving nosepiece

arm

objective lenses

stage

stage clips

coarse-adjustment knob

fine-adjustment knob

light source

base

Procedure

1. **Select a slide and align the slide.** Make sure the low-power objective lens (4x) is in line with or directly over the hole in the stage of the microscope. Select a prepared slide, place it on the stage of your microscope, and secure it with the stage clips.

2. **Align the microscope.** Locate the coarse-adjustment knob on the side of the microscope. Carefully turn the coarse-adjustment knob to lower the stage until the low-power objective lens almost touches the slide. Do not let the objective lens touch the slide.

3. **Adjust the focus of the low-power lens.** Look through the eyepiece. If the slide appears out of focus, use the coarse-adjustment knob to bring the cells into view. Look through the eyepiece as you adjust the focus. When the cells come into view, you can sharpen the focus using the fine-adjustment knob.

4. **Adjust the focus of the high-power lens.** Turn the nosepiece until the high-power objective lens is in line with or directly over the hole in the stage. You will hear it click into place. Use the highest-power objective lens available on your microscope. Refocus using the fine-adjustment knob, and look at the cells using the high-power objective lens.

5. **Observe and record.** On a sheet of paper, draw a circle about 8 cm (3 in.) in diameter for each step of cell division you saw in the video. Each circle will represent a cell on your slide. As you look at each cell, pay attention to the arrangement of the chromosomes. Begin your observations by looking for a cell that closely matches the first sketch you drew when you observed the video. When you find the cell, sketch the arrangement of the chromosomes in the first circle. Find a cell that matches the second arrangement you saw in the video, and sketch that arrangement in the next circle. Make your sketches as accurate as possible.

   Continue looking for cells to match the arrangements you saw in the video. If you have looked over the entire slide and cannot find the arrangement you need, label that circle *Could Not Find* and move on to your next sketch.

6. **Clean up.** When every group member has recorded his or her observations, remove any slides from the microscope. Clean the microscope stage and lenses, and follow your teacher's instructions about how to store the microscope.

Analyze Your Data

Meet with your group and compare your cell observations. Perhaps you and others in your group have different opinions about how many arrangements of chromosomes you observed. Describe how you identified the different arrangements, using evidence from your observations to support your claims. Listen carefully as others present their drawings and their descriptions of the arrangements they saw.

As a group, come to an agreement about the number of different arrangements you observed. Discuss and agree on a description of each arrangement. Be prepared to present your arrangements and descriptions to the class.

Communicate

Share your group's drawings and descriptions with your class. Identify each of the chromosome arrangements in the process of cell division. Describe how you identified the different arrangements, and describe how the chromosomes are arranged in a pattern in each drawing. Pay attention as other groups present their drawings and descriptions. Answer the following questions:

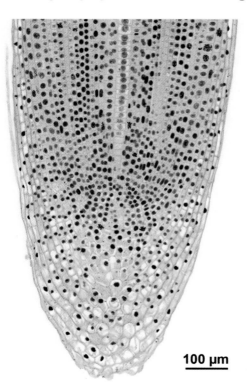

100 μm

Onion root tip

- How did the other groups' steps compare with yours?

- How did others describe what was happening to the chromosomes in their drawings?

As a class, using your observations and drawings, come to an agreement as to how many different arrangements of chromosomes, or steps, there are in cell division.

Observe

Cell-Division Animation

You have observed cell division in two different ways. The video you saw earlier showed the process of a real cell dividing. You then observed cell division by looking at slides of preserved cells. Each cell was frozen in time at a particular point in cell division. However, you may not have been able to find all the chromosome arrangements, or steps, of cell division on your slide. Also, you may have had disagreements about the sequence of the steps of cell division.

You will now have a chance to observe a third example. You will watch an animation of cell division. The cells you saw in the video and on the slide had many chromosomes. The animation shows cell division in an animal cell with only two chromosomes. It is a simulation of the real process. The simulation makes it easier to see the sequence of steps in cell division.

As you watch the animation, match the drawings you made earlier to the steps in the animation. Check to see what steps you might have missed in your drawings, and then revise your drawings and descriptions to make your steps more accurate. Also check that you have accurately drawn all the different chromosome arrangements. Pay attention to where the chromosomes are and how many there are at each step.

Reflect

The animation showed a single cell with two chromosomes. After the cell divided, each of the two cells contained two chromosomes. The chromosomes duplicated, so that the two cells still had the same number of chromosomes as the original single cell.

As a group, come to an agreement on what you saw in the animation. Review the steps of cell division you agreed on earlier. Work with your group to rewrite those steps to make them more accurate.

Then in your drawings, find the point at which the chromosomes duplicated. Describe why it is important that the chromosomes duplicate in this way.

Model and Simulate the Process of Cell Division

The video, prepared slides, animation, and pictures you drew all provided information to help you describe the process of cell division. Now you will design and run a simulation to show how chromosomes transfer to a new cell during cell division. Your drawings and observations will help you plan your simulation. You will use pipe cleaners to model chromosomes and string to model the boundaries of the cells. Your model cells will each have 2 chromosomes, each represented by a different color. You will work in pairs to plan your simulation and to run another pair's simulation plan.

Plan Your Simulation

Materials
- green and red pipe cleaners
- 2 pieces of string, 40 cm long
- 4 paper clips
- blank paper for writing
- scissors

1. Before you write a simulation plan for your model, work with your partner to use the materials (pipe cleaners and string) to walk through each step. You might find this difficult at first, but remember to use your drawings and what you have seen in the video and animation to help you. The goal is to use the materials to replicate the process cells use to make new cells. You should agree about the steps of the simulation and be able to work through them two or three times.

2. Write detailed instructions about how to use the pipe cleaners and string to simulate the process of cell division. Be sure to write your process to accurately describe how your simulation works. Provide enough detail about each step of your simulation so somebody else can follow your directions and accurately simulate how cells divide.

Run Your Simulation

When you have finished writing your simulation plan, you will exchange plans with another pair of classmates and run their simulation. While one pair carries out the simulation, the other pair will observe them and take notes.

When you run a simulation, follow the directions in the simulation plan as exactly as you can.

When you observe the pair of students running your simulation, pay attention to exactly how they are carrying out the steps. Note any differences in what you think you wrote and how the other group is following your plan. Pay attention to the steps the group finds simple and those that are more difficult. Also pay attention to how the simulation is similar to, or different from, the actual cell division you saw in the video.

After both simulations are run, compare your simulation to the pictures you drew and the video and animation of cell division you saw. Identify places where your simulation plan needs to be more specific or needs to be changed.

Conference

Together with the classmates whose simulation you ran, discuss the accuracy of the two simulations. Use your notes to discuss how each simulation matched the drawings you made earlier and your observations from the video, animation, and slides. How did the steps of each simulation differ from what you actually observed in the video, animation, and slides? How could you rewrite each simulation plan so it is more accurate?

Revise the Simulation

As a class, review the steps of cell division. Together, write a simulation plan that accurately shows the process of cell division seen in the video, animation, and slides. Using the materials for the simulation, run the simulation using the new plan. Check to see that the steps make sense. Come to an agreement about the process of cell division and a definition of each step.

Reflect

Consider the process of cell division as you answer these questions.

1. Chromosomes contain the genes that control the traits of an organism. Based on the results of your simulation plan, do you expect new cells to have the same traits or different traits as the original cells? Record your answer, and use evidence from your simulation plan to support your conclusion.

2. When trying to develop a new rice plant, why is it important that you know how chromosomes are transferred to new cells? Use evidence from your investigations to support your answer.

3. What else do you need to learn about cell division and reproduction to answer the *Rice for a Better World Institute's* question about how to make rice plants resistant to caterpillars?

What's the Point?

The cells of organisms die and need to be replaced. Organisms also grow and need more cells. All cells go through a process of cell division to make new cells. This process happens the same way each time. During cell division, one cell divides into two cells. The chromosomes of the original cell are duplicated, and one copy is transferred to the new cell. This process results in two identical cells. They each have the same chromosomes as the original, single cell.

Viewing a complicated process in many different ways helps to understand it. By watching cell division through videos, drawings, animations, and models, you saw the same process in several different ways. You probably understood the process better each time you saw it in a new way.

4.3 Read

How Do Cells Reproduce?

You have seen the process of **cell division** in a number of ways. When the original, or **parent cell**, divides, the result is two identical cells. Scientists call these two cells **daughter cells**. To carry out the functions of the cell, each daughter cell must have the same genetic material as the parent cell. You have observed how the duplicated genetic material was split between the two daughter cells. If this did not happen, the new cells would have only half of the genetic material of the original cell.

You have drawn and observed the steps of cell division, and you will now find out the labels scientists use for these steps. In this section, you will read about the details of the process and use the words scientists use to describe cell division. As you read, think about the different ways you have seen the process of cell division. Make sense of the reading by matching what you are reading to the drawings, video, and model you have seen. It would also be a good idea to match what you are reading to the drawings of the phases, or steps, on the next pages.

cell division: the splitting of a parent cell into two daughter cells.

parent cell: the original cell before it divides.

daughter cells: the two cells that result from cell division.

mitosis: the duplication and splitting apart of chromosomes during cellular division.

phase: a step or a stage in a process.

interphase: the step before mitosis begins, during which a cell prepares to divide.

prophase: the first step of mitosis, during which the genetic material condenses into chromosomes. Each chromosome consists of two identical strands.

Mitosis

Scientists call the duplication and splitting of chromosomes during cell division **mitosis**. You know that scientists look at this process as a series of steps. The steps are called **phases**.

- A cell spends most of its time in **interphase**. In interphase, the cell grows and duplicates its genetic material. If you looked at the cell with a microscope, the genetic material would appear as a dark mass or a tangled mass of threads inside the cell.

- The first phase of mitosis is **prophase**. During prophase, the duplicated genetic material tightens up and forms the coils that you can see as chromosomes.

In the slides, you looked only at plant cells. However, in the animation, you looked at an animal cell. Animal cells have two tiny structures called **centrioles**. During prophase, the two centrioles begin to separate and move to opposite sides of the nucleus, and then the **spindle** forms. You may have noticed the spindle because it looks like a fan made of tiny threads. The spindle helps separate the chromosomes.

Plant cells also have spindles, but the spindles form by themselves.

* Next is **metaphase**. This phase is in the middle of the process. During this phase, the chromosomes line up across the center of the cell. The spindle fibers connect to the center of each chromosome.

* After the spindle fibers have connected to the center of each duplicated chromosome is the step called **anaphase**. During anaphase, the spindle fibers pull apart the strands of the chromosomes and then pull the strands toward opposite ends of the cell.

centrioles: very small structures, found in animal cells, that produce the spindle fibers.

spindle: a structure made up of tiny tubes that attach to the duplicated chromosomes and pull them apart in anaphase of mitosis.

metaphase: the step of mitosis, where the chromosomes line up across the center of the cell.

anaphase: the step of mitosis, during which the strands of the chromosomes are pulled apart by spindle fibers and move toward opposite ends of the cell.

telophase: the step of mitosis when the single strands of chromosomes gather at opposite ends of the cell and no longer are visible as separate chromosomes.

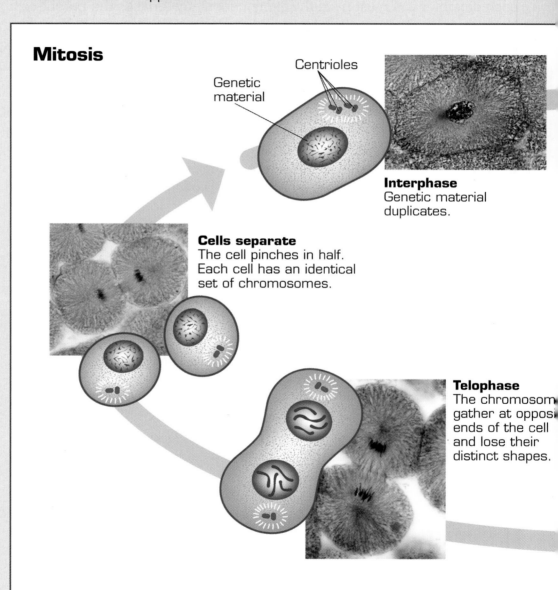

Mitosis

Centrioles

Genetic material

Interphase
Genetic material duplicates.

Cells separate
The cell pinches in half. Each cell has an identical set of chromosomes.

Telophase
The chromosom[e] gather at oppos[ite] ends of the cell and lose their distinct shapes.

- Finally, during **telophase**, the chromosome strands gather at opposite ends of the cell. They bundle into tangled masses of threads again, and they are no longer visible as separate chromosomes.

- The cell then separates into two individual cells. Each of these two daughter cells has an identical set of chromosomes.

Spindle fiber forming

Prophase
The chromosomes condense and become visible. The centrioles separate, and spindle fibers begin to form.

Chromosomes

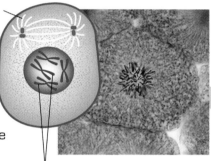

Metaphase
The chromosomes line up across the center of the cell. Each chromosome is connected at its center to a spindle fiber.

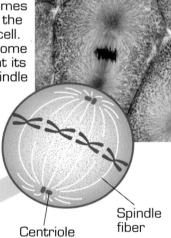

Centriole

Spindle fiber

Anaphase
The duplicated chromosomes separate into individual chromosomes and are moved apart.

Individual Chromosomes

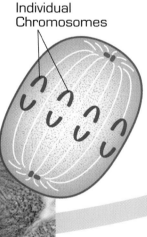

GENETICS

Reflect

1. Review the steps of mitosis you drew and the descriptions you wrote in the previous section after viewing the video, the animation, and the slides. Think about how the steps you recorded match the way scientists define the phases. Revise your drawing and descriptions based on what you now know about mitosis.

2. During interphase, the genetic material in the parent cell duplicates. Why is this step important to the process?

tumor: an abnormal growth of tissue.

Cell Division and Cancer

The normal cells of a multi-celled organism divide when an organism needs to grow or needs to make new cells to repair an injury. Most of the time, mitosis follows the process you observed. When the process works the way it should, the daughter cells are exact copies of the parent cell. Sometimes, however, mitosis does not work well, and the cells it produces include errors. When this happens, the results for the organism can be serious.

Cancer is one disease that can occur as a result of problems during the process of mitosis or other kinds of damage to the genes. Some chemicals can damage genes or cause changes in the instructions a gene contains. Changes in genes and errors during the duplication of chromosomes cause changes in chromosomes. When a changed chromosome divides in mitosis, the new cells inherit the changed genes.

The genes on the chromosomes inside a normal cell tell the cell when to stop dividing and when to die. Scientific evidence suggests that cancer cells lack this important information. The instructions in cancer cells do not tell them when to stop dividing or when to die. Because of this, cancer cells continue to divide and do not die. They also divide more quickly than normal cells.

When cells divide out of control, the cells form **tumors**. Cancer becomes deadly when tumors grow so large that they keep organs from functioning and when cells break off from the tumors and travel to other parts of the body where they grow into new tumors.

Update the *Project Board*

It is now time to update your *Project Board*. Record what you read about mitosis in the *What are we learning?* column. Be sure to include evidence from your reading in the *What is our evidence?* column. Organisms, such as rice and people, have thousands of traits. All the traits are in genes on the chromosomes. There is a lot of variation in these traits. Each new population of organisms is different from the population before it. What else do you need to learn about cells and cell division to understand what is happening inside cells to produce all this variation? Put your new questions in the *What do we need to investigate?* column.

What's the Point?

Mitosis is the name given to the process through which cells divide. Before mitosis begins, during interphase, the genetic material duplicates. In prophase, at the beginning of mitosis, the genetic material condenses into chromosomes. In animal cells, centrioles begin to produce spindle fibers. In plant cells, spindle fibers form on their own. In metaphase, the chromosomes line up across the cell, and spindle fibers attach to the center of each duplicated chromosome. In anaphase, the strands of the chromosomes are pulled apart by the spindle fibers. In telophase, the chromosome strands gather at opposite ends of the cell. After mitosis, the cell divides, and each daughter cell contains chromosomes with the same genes as the parent cell.

Cancer develops from changes in the genes on a chromosome or from a mistake in mitosis. A cell that is damaged from chemicals or viruses can lack the normal instructions about when to stop dividing and when to die. Cancer cells continue to divide and form masses called tumors. Cancer becomes deadly when tumors get so large that other organs cannot function and when cells break off from tumors and grow into new tumors in distant parts of the body.

binary fission: the simplest form of asexual reproduction, in which the parent cell divides in two. Each cell has the same genetic material. Binary fission is used by protozoa, bacteria, and some algae.

asexual reproduction: one parent cell divides into two cells, and each of the new cells has the same genetic material as the parent.

budding: a method of asexual reproduction in which an outgrowth forms on the parent and eventually breaks off and lives independently.

fragmentation: an asexual reproductive process in which an organism breaks into pieces, and each piece grows into a new individual.

vegetative reproduction: an asexual reproductive process in plants in which new cells separate from the parent and form new organisms.

More to Learn

Asexual Reproduction

Each cell in an organism has a complete set of chromosomes with a full set of instructions for the cell to carry on its life processes. Each cell goes through mitosis to divide. Humans and other multi-celled organisms use mitosis to make new cells. The cells divide when the organism needs to grow or to produce new cells to stay healthy.

What is Asexual Reproduction?

Many single-celled organisms reproduce in the same way the cells inside multi-celled organisms divide. They follow the same process you read about and saw in the video and animation. This simplest form of reproduction is called **binary fission**. In binary fission, the chromosomes inside the parent cell duplicate and split apart through mitosis. The parent cell then divides into two daughter cells. Single-celled organisms that use this method of reproduction are bacteria, protozoa, and some algae.

When the reproduction of an organism involves only one parent, scientists call the method of reproduction **asexual reproduction**. The new organism produced by asexual reproduction has the same chromosomes as the parent cell. Generation after generation, the offspring have the same chromosomes as the parent. The offspring inherit all the parent's strengths and weaknesses.

What are Some Other Ways Organisms Reproduce Asexually?

Asexual reproduction can happen in several different ways other than binary fission.

- In **budding**, an outgrowth forms on the parent and eventually breaks off and lives independently.

- In **fragmentation**, an organism breaks into pieces, and each piece grows into a new individual.

- During **vegetative reproduction**, new cells separate from the parent and form new organisms.

- In **spore formation**, an organism forms a special cell called a spore. After the spore is released, it may develop into a new organism.

Methods of Asexual Reproduction

Budding	An outgrowth forms on the parent. Eventually, the outgrowth breaks off and grows into a new individual. Yeast is one organism that reproduces by budding.	
Fragmentation	In fragmentation, an organism may simply break apart into pieces. Each piece then grows into a new organism. This form of asexual reproduction is commonly seen in organisms such as planaria (right) and some worms.	
Vegetative Reproduction	In this form of asexual reproduction, used by plants, new cells separate from the parent and form new organisms. Plants use special structures, such as runners, tubers, bulbs, corms, or rhizomes to produce new plants asexually. Artificial methods of vegetative reproduction include cuttings, grafting, and budding.	
Spore Formation	Spores are special cells, covered by a tough, protective membrane, that are produced by specific organisms. Spores are released from the parent, and when environmental conditions are right, they can develop into a new organism. Some fungi, algae, and protozoa use spore formation as a means of asexual reproduction.	

No matter what method is used, asexual reproduction usually occurs very rapidly. This is why one bacterium can become a billion in a relatively short period of time. As a result of such rapid reproduction, evolution in a population of organisms that reproduces by asexual reproduction can occur much more rapidly than in a population that does not reproduce through asexual reproduction.

Cloning

Cloning is a type of asexual reproduction in which the offspring is genetically identical to its parent. A clone receives all its genetic material from just one parent. Although cloning is often in the news, it is not a new method of reproduction. Cloning has been going on in nature for millions of years. It started when the first bacteria on Earth reproduced, making copies of themselves. When a strawberry plant sends out runners that produce new plants the same as the parent plant, it is cloning. The ability to produce clones of organisms in the laboratory is what is making the news today.

Scientists can clone organisms in laboratories to duplicate organisms that have desirable traits. They insert the genetic material from the organism they want to duplicate into an egg cell of a different individual (instead of the genetic material that is already in the egg cell). Then a small electric current is applied to stimulate mitosis. The cell divides into

spore formation: an asexual reproductive process in which an organism forms a special cell called a spore.

cloning: a type of asexual reproduction in which a group of genetically identical cells are produced by the division of a single cell.

The new plants, produced from the runners of strawberry plants, are clones of the parent plant.

Project-Based Inquiry Science

two cells. Those cells reproduce until they form a small ball of cells, and that ball of cells is placed in a nutrient-rich environment. If everything works the way it should, the ball of cells develops into a new organism that is a replica of the parent cell's genetic material.

Scientists have been developing cloning methods for over a hundred years. However, it was not until a lamb named Dolly was produced through cloning that the subject of cloning became a sensation. On July 5, 1996, after 277 attempts, scientists in Scotland cloned a lamb from a single cell of another adult sheep. It was the first time a scientist had cloned a mammal this way.

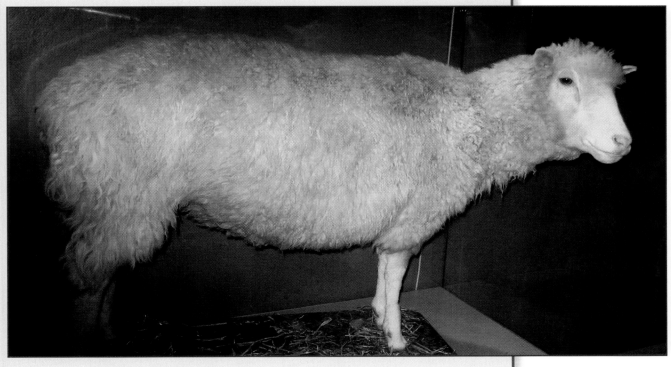

Dolly the sheep was the first mammal successfully cloned from an adult cell. She was born in

There are advantages to asexual reproduction. Only one parent is required for reproduction. Organisms can produce copies of the parent very quickly. However, there are also disadvantages. Recall that the offspring of a single parent inherit both the strengths and weaknesses of the parent. If, for example, a parent lacked immunity to a certain disease and passed this trait on to its offspring, all the other offspring and future generations would then be affected if attacked by that disease.

Project-Based Inquiry

Reflect

1. Rice is a multi-celled organism, and rice does not reproduce asexually. In what situations would you expect that the cells of a rice plant would divide?

2. You read about asexual reproduction to understand how to produce a better rice plant. Why might it be important that you know how genetic material is transferred to new cells through asexual reproduction? Use what you have read to justify your answer.

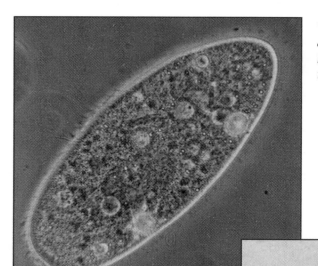

This paramecium is a one-celled organism that uses binary fission to reproduce through asexual reproduction.

This green algae can reproduce by asexual or sexual reproduction.

4.4 Explore

Why Is There so Much Variation Among Organisms?

When your class built the Reeze-ot offspring from one pair of parents, you observed tremendous variation. Every day, you also see that kind of variation when you look around you. People, dogs, and other organisms have traits that are unique to each individual.

But you have also read that the result of asexual reproduction is offspring that have the same chromosomes as the parents. If all organisms made new organisms through asexual reproduction, there would be very little variation in organisms. But we know there is great variation. Therefore, there must be a different process that leads to variation within a species.

The process that accounts for variation is called **sexual reproduction**. Sexual reproduction begins with two **sex cells**, a male sex cell and a female sex cell. (You may sometimes see sex cells referred to as reproductive cells or gametes.) The male sex cells are called **sperm**, and the female sex cells are called **eggs**. In sexual reproduction, a male asex cell nd a female sex cell unite to form one new cell called a **zygote**. The zygote then develops into a new individual organism.

In this section, you will investigate how sex cells reproduce and how sexual reproduction results in variation. As you watch the animation in this section, you will see how the sex cells form, how they reproduce, and what happens to the genetic material inside them.

sex cells: in many organisms, the sex cells are egg cells and sperm cells. Both the egg and sperm are single cells. In humans, the egg is the largest single cell in the body, and the sperm is the smallest.

sperm: the male sex cell; the reproductive cell that carries the male's genetic information to the female's egg; also called the male gamete.

egg: the female sex cell; the reproductive cell that contains the female's genetic information; also called the female gamete.

zygote: a fertilized egg.

Project-Based

Observe the Formation of Sex Cells

The purpose of mitosis in regular body cells is to make new cells to heal and grow. The purpose of sex cells is to make offspring. The purposes are different, so the process used to make new sex cells must be different. In this animation, you will see the process sex cells use to reproduce. You will see what happens to a parent cell when sex cells are formed. The parts of the cell may look very familiar to you. You have seen a lot of these same parts in the mitosis animations and models. As you watch the animation, pay attention to:

- the cell parts that are similar to those in mitosis (nucleus, genetic material [chromosomes], cell boundaries);

- what happens to the genetic material (chromosomes) as the sex cells are formed;

- how the process of the formation of sex cells is similar to the process of mitosis;

- how the process of the formation of sex cells is different from the process of mitosis.

When you watched the video, slides, and animation of mitosis in the previous section, you observed the different stages, or phases, of the process. As you watch the video of the formation of sex cells, pay very close attention to what happens to the number of chromosomes during the process. Do not worry about precisely describing each stage.

Conference

Begin your conference by comparing the animation you just saw to mitosis. You are already familiar with the process of mitosis. Compare the process of mitosis to what you just observed happening during the formation of sex cells. What is happening in the process you just observed that did not happen in mitosis? What about the process is similar to mitosis? How is the outcome of the two processes different? Use a table like the one on the next page to help you organize your discussion.

	Mitosis	Sex cell formation
Process		
Final outcome		
Genetic material		
Number of cells		

Sexual Reproduction

How are Sex Cells Different from Body Cells?

The process by which sex cells are made from a parent cell is called **meiosis**. Meiosis has two parts to it. First, a cell divides into two cells (called meiosis I). Then, each new cell makes two sex cells (called meiosis II). During meiosis, the number of chromosomes in the parent cell is cut in half to produce a sex cell with half the number of chromosomes as in the parent cell.

This is an amazing and important step in the process. Without this step, cells would not be able to re-create the organism they came from. Let's see how it works for humans. The process works the same way in most complex organisms.

Human body cells have 46 chromosomes. If each sex cell, the egg and the sperm, had 46 chromosomes, when the two cells joined, they would produce a zygote with 92 chromosomes (46 + 46). If that process continued into the second generation, the zygote would have 184 chromosomes (92 + 92). Scientists know this does not happen. For a zygote to have the correct number of chromosomes, each sex cell must contain half the number of chromosomes of other body cells. Sex cells in humans have 23 chromosomes. When sex cells join during reproduction, the new cell has 46 chromosomes (23 + 23).

meiosis: cell division that produces sex cells with one half the number of chromosomes found in each body cell.

Observe the Formation of Sex Cells Again

Watch the video again. Pay very careful attention to the chromosomes in each cell. See if you can find the place in the process of meiosis where the chromosomes are divided into two sets. Watch as the cells go from two cells to four. Find the point in the process where chromosomes in these cells divide.

Stop and Think

1. How many chromosomes are there in a human sperm cell or egg cell produced by the process of meiosis?

2. Describe ways that meiosis is different from mitosis. Use your discussion and the animation to help you develop a description.

Observe Fertilization

You have seen the process through which more sperm or egg cells are made. The parent cells provide the genetic material, but each provides only half the chromosomes. To better understand the importance of the process, you will observe an animation of the fertilization of an egg cell by a sperm cell. In the animation, you will see the sperm and the egg fuse together. The result is a fertilized egg, or zygote. The genetic material in a zygote comes from both the male and female parent. The sperm contains genetic material from the male parent, and the egg contains genetic material from the female parent.

As you watch the animation, pay attention to

- when the sperm penetrates the egg;

- the number of sperm there are compared to the number of sperm that penetrate the egg;

- what happens to the sperm after it penetrates the egg;

- the size of the sperm and the size of the egg in the animation.

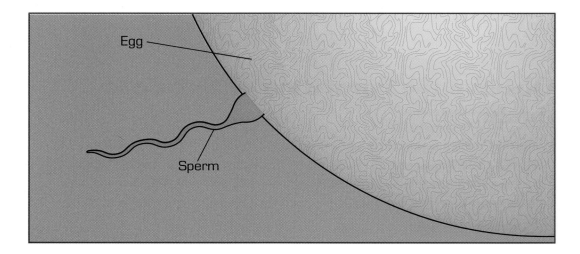

Conference

With your group, describe the animation. What happened in the fertilization that you might not have thought would happen? When the sperm entered the egg, what happened to it, and why do you think this is so?

Complete your conference by discussing the answers to these questions.

- How is the outcome of fertilization different from the formation of sex cells?

- How are the processes of fertilization and the formation of sex cells dependent upon one another?

Communicate

Discuss the process of fertilization with your class. Listen carefully as each group presents their descriptions of fertilization. Try to come to a conclusion about how the chromosomes of the sperm and egg combine to produce the correct number of chromosome pairs in the fertilized egg.

Sexual Reproduction

Meiosis

As a result of meiosis, human sperm and egg cells have only 23 single chromosomes, compared with 23 pairs of chromosomes (46 chromosomes) in body cells. When an egg with 23 chromosomes and a sperm with 23 chromosomes fuse during fertilization, the fertilized egg has 46 chromosomes in 23 pairs, exactly the same number as in body cells.

In fertilization, the sperm from the male parent and the egg from the female parent combine to produce a zygote. Half the genetic material in the zygote comes from the female parent, and half comes from the male. This process, the combination of meiosis and fertilization, can account for a great amount of variation in a population.

Meiosis and Variation

The biggest difference between sexual reproduction and asexual reproduction is that the offspring from sexual reproduction receive chromosomes from two different cells instead of one cell. They receive one chromosome in each chromosome pair from each cell.

GENETICS

When you built the offspring Reeze-ots from one pair of parent Reeze-ots, each of the offspring had a unique combination of traits. The variation you observed resulted from each offspring receiving half of its genetic information from one Reeze-ot parent and half from the other Reeze-ot parent. The offspring produced by sexual reproduction are genetically different from the parent or parents, and each offspring has a unique combination of traits.

As a result of meiosis, the number of possible chromosome combinations in sex cells and a fertilized egg is very large. With 23 chromosome pairs, there are over 8 million possible chromosome arrangements. Rice cells have only 12 pairs of chromosomes. Species that have a larger number of chromosome pairs usually show more variation among individuals than species with a smaller number of chromosome pairs. But even 12 pairs of chromosomes are enough for a great amount of variation.

Meiosis

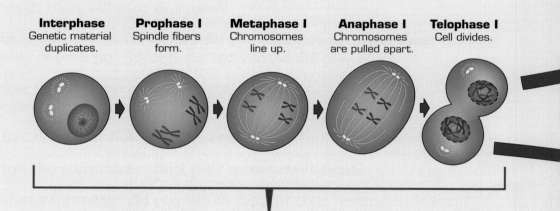

| **Interphase** Genetic material duplicates. | **Prophase I** Spindle fibers form. | **Metaphase I** Chromosomes line up. | **Anaphase I** Chromosomes are pulled apart. | **Telophase I** Cell divides. |

Meiosis I

The stages of meiosis I are exactly the same as mitosis. One parent cell divides into two daughter cells. Each daughter cell has the same genetic material as the parent cell.

Male or Female? The Sex Chromosomes

The traits of all organisms, including being male or female, are controlled by genetic information in the organisms' cells. One pair of chromosomes determines if an organism will be male or female. These chromosomes are called **sex chromosomes**. In humans, these two sex chromosomes are called *X* and *Y* chromosomes. The *X* and *Y* chromosomes have different genes. The *X* chromosome contains genes that determine the characteristics of females, and the *Y* chromosome has genes that determine the characteristics of males.

All male body cells have one *X* and one *Y* chromosome. After meiosis, half of the sperm cells have an *X* chromosome and half have a *Y* chromosome. All female body cells have two *X* chromosomes. After meiosis, all the eggs have one *X* chromosome.

sex chromosomes: the chromosomes that determine the sex of an individual. In humans, these chromosomes are known as X and Y.

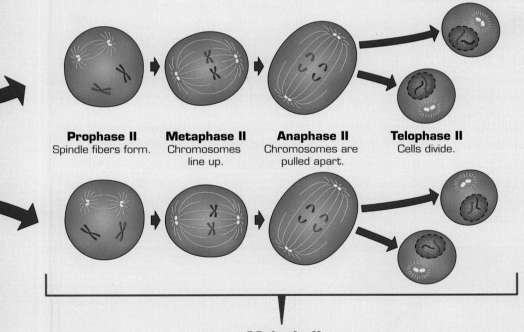

Prophase II
Spindle fibers form.

Metaphase II
Chromosomes line up.

Anaphase II
Chromosomes are pulled apart.

Telophase II
Cells divide.

Meiosis II

The two daughter cells from meiosis I go through meiosis II. Meiosis II has the same stages as meiosis I except that it has no Interphase, so there is no duplication of the genetic material in the daughter cells. Each daughter cell then divides into two additional daughter cells. The result of meiosis II is four daughter cells, each daughter cell with half the genetic material of the original parent cell.

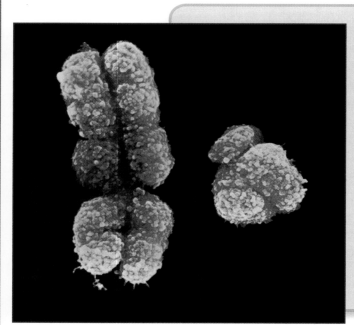

During fertilization, if both the egg and sperm have an *X* chromosome, the zygote will have two *X* sex chromosomes (*XX*). A zygote with an *XX* pair of sex chromosomes will develop into a female. If the sperm has a *Y* chromosome, when it combines with an egg, which always has an *X* chromosome, the zygote will have one *X* and one *Y* sex chromosome (*XY*). A zygote with an *X* and a *Y* chromosome will develop into a male.

The X chromosome (on the left) contains genes that determine female characteristics. The much smaller Y chromosome (on the right) contains genes that determine male characteristics.

Reflect

1. Think back on the Reeze-ot activity when you built a classroom full of Reeze-ots from one pair of parents. Using what you know about sexual reproduction, describe why the variation happened.

2. There were no male or female Reeze-ots because the trait for sex was not included in the list. Imagine if the sex trait had been included. Using the *XY* label for males and the *XX* label for females, create a Punnett square for the trait *sex*. What percent of the population do you think would probably be male if the Reeze-ots had sex chromosomes? What percent of the population would probably be female?

		Male	
		X	Y
Female	X		
	X		

Explain

You will now make a claim about why there is so much variation among organisms that reproduce sexually. Then, using a *Create Your Explanation* page, you will develop an explanation of your claim and support it with evidence. Your *Create Your Explanation* page helps you make sure your explanation connects your claim to your evidence and science knowledge. Your claim should be about why there is so much variation among individuals produced through sexual reproduction. Your evidence comes from the pictures you have drawn. Your science knowledge comes from your reading and exploration. You may also have some science knowledge from your own experiences or from other readings. After you have recorded your claim, evidence, and science knowledge, write a statement using your evidence and science knowledge that supports your claim and tells why it is valid. This is your explanation. Make your explanation a statement about the process that results in your claim.

Communicate

Share Your Explanation

Share your group's claim and explanation with the class. Tell the class what makes your claim accurate based on your evidence and science knowledge. Pay attention to how the other groups have supported their claims with science knowledge. Ask questions or make suggestions if you think a group's claim is not as accurate as it could be, if the group has not supported their claim well enough with evidence and science knowledge, or if a group has not described the process of meiosis well.

Create Your Explanation

Name:_____ Date:_____

Use this page to explain the lesson of your recent investigations.

Write a brief summary of the results from your investigation. You will use this summary to help you write your Explanation.

Claim – a statement of what you understand or a conclusion that you have reached from an investigation or a set of investigations.

Evidence – data collected during investigations and trends in that data.

Science knowledge – knowledge about how things work. You may have learned this through reading, talking to an expert, discussion, or other experiences.

Write your Explanation using the *Claim*, *Evidence*, and *Science knowledge*.

Revise Your Explanation

As a class, come to an agreement on an explanation of why there is so much variation in poputations of organisms that reproduce sexually.

Reflect

Answer the following questions and discuss your answers with your class.

1. How do the processes of meiosis and fertilization help explain the differences in traits among individuals?

2. A normal rice cell has 12 pairs of chromosomes. How many chromosomes do you think each of the sex cells contains? How many chromosomes do you think the zygote of a rice plant will contain?

3. A zygote is a single cell that grows into an individual organism. What type of cell division will a zygote undergo as it grows and develops? Why is it important that the cell uses that type of reproduction? Support your ideas with evidence.

4. How will your knowledge of the process of sexual reproduction help you develop a new rice plant?

Update the Project Board

It is time to update your *Project Board*. Record your new knowledge, especially about meiosis, fertilization, and variation, from your exploration and reading in the *What are we learning?* column. Record the evidence to support your ideas in the *What is our evidence?* column. You should be developing some additional questions based on what you have explored and read about. Update the *What do we need to investigate?* column with these new questions and ideas for investigations. You will have a chance to answer these questions later in this Unit. As the class *Project Board* is updated, do not forget to revise your personal *Project Board*.

What's the Point?

In sexual reproduction, male and female sex cells join together in fertilization. The offspring receives half of its chromosomes from one sex cell and half from the other sex cell.

To produce offspring with the correct number of chromosomes, sex cells must have half the chromosomes of body cells. The halving of the chromosomes occurs during the process of meiosis. This process is one reason why sexual reproduction results in differences in traits between parent and offspring and among individuals in a population. Each offspring has some traits of the male sex cell and some traits of the female sex cell and therefore possesses a unique combination of traits. There are millions of possible chromosome combinations in humans. This explains how sexual reproduction increases genetic diversity.

More to Learn

Genetic Disorders

Down syndrome: a genetic disorder that results when chromosome number 21 fails to separate during meiosis.

In the previous section, you read about cancer, which results from abnormalities in the genetic material in some cells of the body. Many genetic disorders are caused by abnormalities in cell material in a zygote. When this happens, all the cells in the new organism have that same abnormality. Some abnormalities in zygotes result from errors that occur during the process of meiosis. **Down syndrome** is one of the most common genetic disorders. Down syndrome results from the failure of one single chromosome (number 21) to separate during meiosis. A child born with Down syndrome has three copies of chromosome 21. Because of this error, he or she has 47 chromosomes rather than the normal 46. Down syndrome causes mental challenges, distinctive physical characteristics, and poor muscle tone in infants. People with this condition often have more heart defects, digestive problems, hearing loss, and leukemia than the general population. Since this is a genetic birth disorder, scientists cannot cure Down syndrome. They can diagnose the disorder before birth and treat the conditions that result from Down syndrome.

4.5 Read

What is the Genetic Material in Genes and Chromosomes?

Your journey to learning about genetics started with observing expressed traits. You tried curling your tongue, and you examined your ear lobes. When you needed information about cells, you used a microscope to see how they divided, and then you watched videos of cell reproduction. You know that alleles that make up genes on chromosomes determine what traits an organism will have, you know the processes organisms use to reproduce and pass on their traits, and you know the processes cells use to reproduce so organisms can grow and mature. But you do not know yet exactly what alleles, genes, and chromosomes are made of. To give the *RBWI* scientists advice about selecting seeds that will produce plants resistant to caterpillars, you need to know more about what the genetic material in cells is made of and how that material tells an organism what traits to have.

The genetic material that makes up the chromosomes and genes of all cells, both plant and animal, is called **DNA** (deoxyribonucleicacid). The DNA in a rice plant carries instructions for producing a rice plant. The DNA in a person contains instructions for producing a person. Variations in the DNA determine the traits a particular rice plant or person will have.

DNA: the genetic material in a cell that stores and transmits genetic information from one generation to the next.

DNA is a complicated chemical, and the discovery of DNA took over 100 years. Scientists have been able to discover how DNA works only because they have built on the work of earlier scientists. Understanding DNA has taken so long, too, because DNA is very small, too small to be seen through the kind of microscope you have been using. Scientists had to wait to make discoveries about DNA, until they had the tools available to allow them to examine a molecule as small as DNA.

Although DNA is incredibly small, it is very powerful. DNA is found in the nucleus of every cell in a plant or animal. DNA in cells' nuclei (plural of nucleus) contains instructions for everything cells do. It contains all the information that is needed for the cell to survive and reproduce. The information in DNA also describes everything about how an organism works. In animals, the DNA also carries all the genetic information for producing offspring. Through DNA, traits get passed from one generation

heredity or **inheritance:** the passing along of information from one generation to the next.

membrane: a covering around a cell that allows materials to flow in and out.

to the next. This movement of information from one generation to the next is called **heredity**, or **inheritance**.

In this section, you will read about DNA, the genetic material in genes and chromosomes. As you read, pay careful attention to how many people have worked to make discoveries about DNA and how they have built on the work of others. Also, pay attention to how discoveries about DNA have helped scientists increase their understanding of genetics and genetic technology over time.

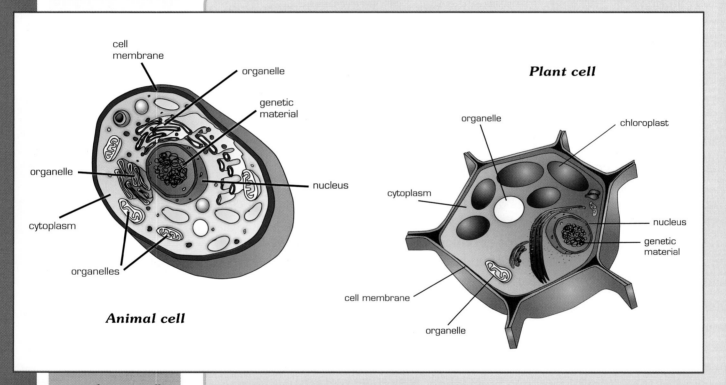

cell membrane
organelle
genetic material
organelle
cytoplasm
nucleus
organelles

Animal cell

Plant cell

organelle
chloroplast
cytoplasm
nucleus
genetic material
cell membrane
organelle

cytoplasm: a jelly-like substance that fills up the space inside the cell.

organelles: structures in the cytoplasm that perform special functions in the cell.

The main parts of a cell are the **membrane**, which surrounds and protects the cell, the **cytoplasm**, a jelly-like substance that fills up the space inside the cell, **organelles**, structures in the cytoplasm that perform special functions, and the **nucleus**, which contains DNA. Plant cells also have a rigid, protective **cell wall** that covers the membrane.

How Was DNA Discovered?

The story of the discovery of DNA begins about 140 years ago. In 1869, a Swiss scientist, Friedrich Miescher, first discovered DNA. His discovery is interesting because he was not actually looking for DNA. He was interested in learning more about **white blood cells**.

Miescher knew that he could find white blood cells on used bandages, so he collected used bandages from a hospital and took them to his laboratory. As he looked at the cells he collected, he found a substance in the nucleus of cells that nobody had ever seen before. He called this substance "**nuclein**." He found nuclein in every single bandage he collected. At that time, many scientists were trying to understand how cells worked, so Miescher reported his discovery of nuclein even though he did not know the job nuclein played in the cells.

Miescher was curious about how cells work and specifically what nuclein was for and what it was made of, so he continued studying nuclein. He looked at many different kinds of cells, and he found nuclein in all the animal and plant cells he looked at. But he never figured out how cells used nuclein.

Miescher's discovery of nuclein came only two years after Mendel completed his experiments with peas, but nobody knew that the two discoveries were connected until many years later. It was not until 1944, 75 years after Miescher's discovery of nuclein, that scientists began to suspect that nuclein was responsible for heredity.

When scientists identified the chemicals that made up nuclein, they changed its name to DNA, an abbreviation of the names of its chemicals. But not many scientists studied DNA, because it was too small to see through the microscopes of that time. DNA had to be studied without being able to look at it. This made studying DNA very difficult. For many years, only a few scientists were investigating DNA. In the 1950s, 90 years after Miescher's discovery, one of those scientists, Rosalind Franklin, completed experiments that made it possible for scientists to identify the atoms that make up DNA without being able to look at DNA itself. She collected data about DNA by using x-rays. She and others used x-rays to collect many bits of information about the atoms that DNA is made of.

nucleus: the control center of the cell. It contains DNA.

cell wall: a rigid, protective covering in plant cells.

white blood cells: blood cells that protect the body against diseases.

nuclein: the name Friedrich Miescher gave to the material in cells that we now call DNA.

Friedrich Miescher was from Switzerland and was the first to discover DNA. He discovered DNA in 1869.

Other scientists at the time were studying molecular structure — the kinds of structures molecules can have. One of those people was Linus Pauling. His discoveries of the structures of molecules also contributed to discovering the structure of DNA. His research was about the different ways molecules can attach to each other.

Rosalind Franklin used x-rays to study the structure of DNA. Watson, Crick, and Wilkins used the results of her studies to determine the structure of DNA.

How can Scientists "see" DNA?

You have used a microscope to see chromosomes inside a cell nucleus. Chromosomes are too small to see without a microscope, but you can see them when a microscope magnifies them 100 times. The molecule that makes up the chromosomes and genes of all cells, DNA (deoxyribonucleic acid), is much smaller. DNA is so small that scientists who first investigated the molecule and its role in organisms could not see it, even with a microscope. Because they could not see DNA, they had to design experiments where they could see the outcome of changes in DNA. Rosalind Franklin used x-rays to identify changes in DNA, but she was never able to see DNA itself. James Watson, Francis Crick, and Maurice Wilkins, scientists who discovered the structure of DNA, figured out the structure of DNA without ever seeing DNA through a microscope.

Scientists still cannot directly see DNA through a microscope. However, a very powerful kind of microscope called a **transmission electron microscope** can produce an image of DNA on a screen. It is not important to know exactly how this kind of microscope works, but it is important to realize how this new kind of microscope has helped scientists understand DNA and genetics. Using this tool, scientists have been able to capture images of very small molecules, such as DNA.

Linus Pauling provided his research results to assist Watson, Crick, and Wilkins in determining the structure of DNA. Pauling continued his career and eventually earned two Nobel Prizes.

In 1953, Crick, Watson, and Wilkins identified the structure of DNA molecules using the x-ray data Rosalind Franklin and others had collected and the discoveries of molecular structure that Pauling had made. An important method they used in making their discovery was modeling. They used Franklin's x-ray data to imagine possible structures for DNA. Then, using sticks and balls, they built a model of each of the structures they imagined. They used the stick-and-ball structures to help them better imagine what the x-ray data was telling them. They built models that matched each of the small puzzle pieces that had been found through x-rays. Then they worked to put those small models together with each other to create big structures, making sure that their big structures followed the rules Pauling had discovered.

Crick, Watson, and Wilkins determined that DNA is structured as a "**double helix**"—two long strands of beads wound around each other like a twisted ladder. Pictures you have seen of the double helix usually show it flat (two-dimensional), but really the molecule is three-dimensional. Imagine a long ladder. Now imagine that someone held the ladder firmly at the bottom and twisted the top so that the whole ladder began to spiral. DNA molecules look like this twisted ladder.

transmission electron microscope: a microscope that uses electrons instead of light to produce images of very small objects.

double helix: a twisted ladder structure.

bases: chemical structures that make up the rungs of a DNA ladder.

Francis Crick, James Watson (both pictured), and Maurice Wilkins succeeded in solving the puzzle of what DNA molecules look like.

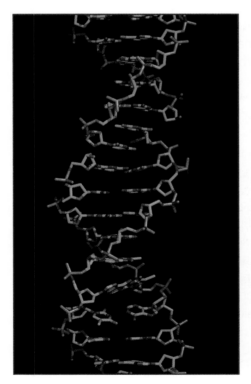

A model of DNA.

The rungs of the DNA ladder are made up of **bases**. The bases are made from four different chemicals: A is Adenine, T is Thymine, C is Cytosine, and G is Guanine. Every strand of DNA is a sequence of these four chemicals. The ladder shape of DNA is formed when two long strands of DNA link up with each other. Strands of DNA link in predictable ways to form rungs of the DNA ladder: A always links to T, and G always links to C.

You can see how the four bases connect to each other in the diagram of the DNA molecule. You do not need to remember the names of these chemicals; just think of them as matching parts that make up the rungs of the DNA ladder. Every DNA molecule is made from a different arrangement of these four bases.

Stop and Think

1. Several scientists were described in this section. Imagine how Watson, Crick, and Wilkins must have felt when they finally were able to build a model of DNA. Describe the feelings you think these scientists had.

2. Each scientist in this story used the work of others to make new discoveries. Record how Watson, Crick, and Wilkins used the work of others.

What is the Role of DNA?

Today, scientists know that DNA is found in the cells of all living things—plants, animals, and bacteria. Scientists know that DNA controls everything cells do and how organisms function. DNA is also responsible for passing on genes to the next generation. Along every cell's chromosomes are the genes that contain the information that controls the traits of organisms. The alleles that make up each gene are made up of a different arrangement of DNA molecules. The actual sequence that makes up each allele could be up to several thousand DNA molecules long.

All cells in an organism contain DNA molecules, but the arrangement of those molecules is different for different traits. All the traits of rice grains, such as size or amount of starch, are controlled by genes. Different types of rice plants, which contain different genes, each have a different arrangement of DNA. These different arrangements result in different traits. Small variations in the same genes among plants of the same type of rice result in small variations among the grains.

In animals, DNA also controls the traits of each individual. For example, whether your ear lobes are attached or detached is controlled by a gene in your cells. The gene for ear lobes occurs in two slightly different forms, or alleles, one for attached ear lobes and one for detached ear lobes. The DNA sequence for each form, or allele, is different. Depending on which DNA sequence you carry in your cells, your ear lobe will be either attached or detached.

The Human Genome Project

Because DNA is essential for life, scientists are studying the DNA of many different organisms. They want to understand exactly how DNA controls the way organisms function and look. The complete set of genes of any organism is called its **genome**.

genome: the complete set of genes of an organism.

gene mapping: finding the location of every gene on a chromosome.

To better understand how DNA works, scientists are studying the genomes of many different organisms. When scientists look at genes, they analyze the sequence of bases along a part of the DNA ladder. Scientists have found the locations of different genes on human chromosomes and on the chromosomes of other organisms. Finding the location of each gene is called **gene mapping**.

Perhaps the most exciting gene-mapping project scientists have done was the Human Genome Project. The Human Genome Project was a large international effort to map the genome of humans. The project was a collaboration of scientists from many different countries, including the United States, the United Kingdom, China, Australia, France, Germany, and Japan. It was one of the largest scientific projects ever completed.

The Human Genome Project began in 1990, and it was originally led by James Watson, the same man who helped discover the structure of the DNA molecule. Before the project began, humans had very little information about their own DNA. The main goal of the project was to develop a map of all human DNA.

How does the Human Genome Project Support Evolution?

In his *Theory of Natural Selection*, Charles Darwin claimed that the environment causes the change to traits in a population over time. Even though Darwin had no knowledge of genes, he hypothesized that all organisms came from common ancestors and then branched off because of differences in the environment. He looked at fossils and found many species that lived a long time ago and became extinct. These species were similar to but different from species he saw in his own time. Darwin used these examples as evidence for his theory of natural selection.

Fossil fish that became extinct long ago.

Now that scientists are able to map genes of different animals, they are finding genetic similarities among species that provide additional evidence for Darwin's theory of common ancestry. For example, mapping has allowed scientists to discover that humans and chimpanzees share from 96% to 99% of their genes. The more genes organisms share, the closer the relationship between them. The DNA of chimpanzees shows that these apes are more closely related to humans than any other organism. But even organisms that are very different from humans have some DNA in common. Humans share about 81% of their genes with mice and 75% of their genes with a small worm.

Another goal of the Human Genome Project was to use the genome map to uncover DNA variations in individuals. This information may lead to new ways to diagnose, treat, and prevent genetic diseases and defects. The information from the map will also be used to study human evolution.

The process of mapping the human genome required the work of many scientists. A leader in this effort was Francis Collins, who had also investigated cystic fibrosis and Huntington's Disease. Collins followed Watson as the Director of the Human Genome Project in 1993.

Scientists had to determine the structure of 20,000 to 25,000 genes. They also had to work out how the bases in DNA work. Scientists developed ways to store all the information and to share the information with others. Scientists also considered many legal and social questions that surrounded the project.

Mapping the human genome required inventing new techniques and developing new machines. The new tools used to map the genome helped scientists advance the project. In June of 2000, President Bill Clinton and the directors of the Human Genome Project announced that they had developed a draft copy of the human genome. The draft copy was shared with the rest of the world through scientific journals.

Scientists continued their work on uncovering the human genome, and by 2006, they announced that they had mapped each of the 23 human chromosome pairs. Many of the scientists involved in the Human Genome Project continue working in genetics. They are using the information from the mapping project to help prevent diseases or to make changes in how diseases are treated. The Human Genome Project is one example of how scientists, working together, can solve tremendous puzzles and make the information available to everyone.

The Breast-cancer Gene

Mapping the human genome has led to knowledge of individual genes and what they do. One disease that has been linked to genes is breast cancer. In 1994, scientists found two genes—on chromosomes 17 and 13—that are linked to breast cancer. A person can inherit a changed form of these genes. Having the changed genes does not mean that a person will get breast cancer, but it does indicate an increased risk of getting breast cancer. That means somebody with the changed form of the genes is more likely to get breast cancer than someone who

Former First Lady Betty Ford suffered from breast cancer. Before she had breast cancer, people were not aware of how many women suffered from this disease.

has the normal form of the genes. However, genes are not the only reason someone may have increased risk of breast cancer. Family history and the environment can also increase a person's risk.

People can now be tested to see if they have the changed form of these genes. They can find out if they are at higher risk. Using knowledge from the Human Genome Project, some day scientists may develop a vaccine for breast cancer.

Many celebrities have had breast cancer and have used their popularity to draw attention to the need for research to treat and eliminate breast cancer. One of the first of these women was former First Lady Betty Ford. She was diagnosed with breast cancer in the 1970s. At that time, people did not talk much about breast cancer. She spoke openly about her disease and how it was treated. Her breast cancer was cured, giving hope to many women. Today, early detection and specific treatment plans for breast cancer have made the disease much more curable.

Reflect

Use your science knowledge from this reading section to answer these questions. Think carefully about the answers, and be ready to support your ideas with information from the text.

1. What were the goals of the Human Genome Project? Which goal is the most important?

2. Scientific projects like the Human Genome Project are very expensive. They also require a lot of people to complete. Describe why you think this project may or may not have been important.

3. Select one of these questions. Record your answer and be prepared to support your answer with evidence.

   • How can gene mapping help scientists trying to cure diseases?

   • How can gene mapping help scientists in developing new crops?

   • How can gene mapping help ordinary people and their families?

4. How can what you now know about DNA and genes help you develop your recommendations for developing a new rice plant?

5. How can knowledge of genes, their functions and their locations, help scientists develop a new rice plant?

Update the *Project Board*

Record what you now know about DNA and gene mapping in the *What are we learning?* column. You read many things about DNA — how it was discovered, its structure, and its function. Record everything in the *What are we learning?* column that you think will be helpful in addressing the *Big Challenge* or answering the *Big Question*. Remember that the *Big Question* is *How can knowledge of genetics help feed the world?* Record evidence for anything you add to the *What are we learning?* column in the *What is our evidence?* column. You may have more questions now about the role of genetics in disease. You can add these questions to the *Project Board* in the *What do we need to investigate?* column.

What's the Point?

Chromosomes are made from very small molecules called DNA. This molecule is shaped like a double-helix, is located in the nucleus of cells, and contains all of the instructions needed to make a new organism. Different arrangements of the four bases of DNA make up the different genes on the chromosomes. These genes carry the information for traits.

Because DNA is essential for life, scientists want to understand how it controls the way organisms function and look. As scientists map the location of each gene, they find out more about how plants, animals, and humans are related to each other. Scientists have spent a great deal of time doing gene mapping. The Human Genome Project took 16 years, and scientists continue working to determine the function of all the genes located on human DNA. Knowing the functions of genes may help prevent or cure genetic diseases.

More to Learn

How Can Scientists Change Traits?

Scientists have made many genetic discoveries since Darwin's time. They have looked closely at the genes in animals and plants. They have learned about the impact of selection pressure from people and the environment. They have also investigated how DNA works and mapped several organisms' DNA. This has led to the next scientific step, changing the DNA of different organisms.

A letter from the Philippines

In my last letter I told you about how rice is grown and how important rice is to my family. I wanted to tell you more about how important rice is as a food in my country. You know that everyone here eats rice several times a day. My family eats white rice. We also eat lots of fish, meat, and vegetables to be sure we get good nutrition.

We recently learned that brown rice is more nutritious than white rice. Brown rice is the actual seeds that grow on the rice plants. This rice is very nutritious because many vitamins and minerals are in the parts that are removed to make white rice. White rice is made by removing the outer covering of the rice grain and polishing it. Vitamin A is one of the vitamins in the endosperm of the rice, and that part is removed when the rice is polished. Vitamin A is a very important vitamin. It is used in the body to improve your eyesight.

Some people do not eat as much fish, meat, and vegetables as we eat. Because they eat a diet of mostly white rice, they may not be getting the vitamins and minerals they need.

Thank you.

Amihan

In the last 20 years, scientists have discovered ways to take DNA from the cells of one organism, cut it into smaller pieces, and attach it to the DNA of another organism. Scientists carefully select pieces of DNA from one organism so the changed organism has the traits needed to solve a problem. When scientists change genes like this, it is called **genetic engineering**.

When people need an organism to have a trait it does not have, like rice having more vitamin A, being resistant to a particular insect pest, or surviving in a drought, **genetic engineers** look for ways to replace a trait of the organism with another trait that is more beneficial. The scientists are not just cross-pollinating or selecting the right seeds; they are making new organisms that have the desired trait in their DNA. This process is called **genetic modification**.

What Does It Mean to Genetically Modify an Organism?

As you read in the letter from Amihan, people who eat a diet of mostly white rice are finding that they are not getting the right combination of vitamins and minerals. It is not possible to improve the amount of vitamin A in rice by choosing traits that already exist in seeds. Instead, genetic engineers use gene mapping to find new ways to increase the amount of vitamin A in rice.

Scientists take the gene for more vitamin A from another organism and insert it into the rice DNA. When they do this, they permanently change the DNA of that rice plant, and the plant produces seeds that contain more vitamin A. These new seeds have better nutrition than the old seeds.

Genetic engineering has been used to improve organisms for use by people. For example, scientists have engineered a rice plant with an increased resistance to bacterial blight, a rice disease. This means more rice can be grown. Scientists can also take strawberries, which are normally soft when ripe, and turn off this trait. This helps strawberry farmers because the firmer strawberries are easier to ship.

genetic engineering: changing the genes of organisms.

genetic engineers: scientists who change the genes of organisms.

genetically modified: used to describe food or organisms that have been genetically engineered by inserting a gene from another organism.

Cows can be genetically engineered to produce milk with increased proteins that benefits people.

Many of the fruits and vegetables you eat are genetically modified. You probably eat genetically modified foods without knowing it. Tomatoes, corn, and some fruits have been genetically modified to have traits that make the plants grow more fruit, live longer, resist pests, or produce fruits or vegetables that are easier to store or move from place to place.

Some medications are also made through genetic engineering. For example, people with diabetes regularly use insulin to maintain their health. Today, insulin is produced by inserting human genes into bacteria. Insulin is far easier and less expensive to produce today than it was many years ago.

The gene that makes strawberries soften can be turned off. The firmer berries are easier to ship.

Genetic engineering works because all cells in living organisms function in about the same way, and they all contain DNA. The similarities among genes allow one gene to be taken from a **donor organism**, inserted into another organism, and then function in the cells of the new organism.

Bt-corn has been engineered to resist a corn pest.

Insulin, a life-saving chemical for many people, is made by inserting human genes into bacteria.

The Story of Bt-Corn

One example of a genetically modified crop is Bt-corn. An insect called the European corn borer was eating corn grown in the United States. This insect had somehow found its way from Europe to the U. S., probably in a cargo ship that was bringing something else here. Because the insect was not native to the U. S., the corn had no resistance to the borer. Farmers could not sell ears of corn that had been partially eaten by a corn borer.

Regular corn (on right) and attacked by corn borer (on left).

Farmers and genetic engineers examined all the kinds of corn seed they knew of to find one resistant to the European corn borer. The corn also had to grow well in the U. S. They could not find one. But they knew of a soil bacterium (a kind of bacteria that lives in soil) that produces a chemical that would kill the borers. They chose the soil bacterium as a donor organism for the trait to kill the borer. In a lab, they extracted the gene for killing borers from some of the bacteria, and they inserted the gene into chromosomes of corn. This produced a new type of corn. This corn's pollen has a chemical in it that kills the **larvae** of the corn borer before they have a chance to grow and eat the corn. They then used the corn seeds they had produced this way to grow more corn and generate more seeds. They grew corn from the seeds, and the seeds produced from those seeds, and so on, until they had enough of the genetically modified corn for all the farmers who grow corn to use in their fields.

This genetically modified type of corn was named Bt-corn. It has become the standard corn grown in the United States. Because Bt-corn is resistant to corn borers, farmers who grow it do not have to spray insecticide for the corn borer on their fields.

Are There Problems with Growing Bt-corn?

When any new crop is developed, scientists carry out experiments to find out if it is safe—for people, other animals, and the environment. Bt-corn was developed to prevent corn borers from eating the corn crop. The pollen of the Bt-corn kills the larvae of the corn borer before they can fully develop and eat the corn.

This is good for corn, but scientists and environmentalists worried that the pollen might also kill the larvae of other insects. They were particularly worried that Bt-corn might be harmful to Monarch butterflies. Because corn borers are very similar to caterpillars that turn into Monarch butterflies, scientists were worried that the pollen from the Bt-corn would also harm Monarch butterfly larvae. Scientists also wondered if the new corn would harm populations of adult butterflies. If Monarch butterflies disappeared because their larvae were killed by the Bt-corn, or if adult butterflies were harmed and could not pollinate plants, many different parts of the ecosystem would be affected.

donor organism: an organism that supplies a gene for another organism.

larvae (singular: larva): the first stage of an insect after it leaves the egg.

Monarch butterfly

To answer their question, scientists set up a laboratory experiment. They knew that butterflies feed on milkweed plants, so they dusted some milkweed plants with pollen from the Bt-corn and left some milkweed plants without the extra pollen. Then they allowed some butterflies to feed on Bt-corn-dusted milkweed and some butterflies to feed on the normal milkweed. The results showed that the butterflies that ate the Bt-corn pollen were more likely to die than butterflies that fed only on the normal milkweed.

Monarch caterpillar

These results caused a lot of panic among scientists and environmentalists. They were concerned that Bt-corn would harm the environment. Some people were also concerned that Bt-corn could harm people. Newspapers, television, and radio produced many stories about the possible harmful effects of Bt-corn.

But scientists were not completely sure about the results of the experiment they had done. They continued to ask many questions about the procedure and data analysis. They were particularly unsure about whether the results they got in the lab would be the same in the natural environment. To address their concerns, they moved the experiment out of the laboratory and observed the effect of the Bt-corn pollen on butterflies in nature. When they did this, they found that the survival rates of the butterflies that were exposed to the Bt-corn pollen were similar to the survival rates of the butterflies exposed only to normal milkweed. Some scientists are still concerned about possible effects of Bt-corn on the environment.

How Safe is Genetic Engineering?

Nature is very complicated, and we cannot predict all of the consequences of changing the traits of organisms. New genetically modified food products are being closely watched to see how the changes in organisms are affecting the environment, including people and wildlife.

Some people have concerns about the safety of genetic engineering. Others see it as opening up great opportunities in agriculture, food, and medicine.

The safety of the new food products, for people and for the environment, is being carefully studied. In the United States, several governmental agencies, including the FDA (Food and Drug Administration), USDA (U. S. Department of Agriculture), and EPA (Environmental Protection Agency), run experiments to monitor the safety of genetically modified foods. Many states have laws that require food producers to label genetically modified food so people can know what they are purchasing at the store. However, since genetic engineering is a new science, all the possible effects of this technology are not known.

There are still many debates about the effects of these new foods. Some people see genetic engineering as opening up great opportunities in agriculture, food, and medicine. Other people think that because we still cannot predict all the effects of genetic engineering, we should not use genetic engineering because it might be harmful to humans and the environment. These people argue that we should find other solutions to our food and medical problems instead of genetic modification. They also worry that the people making decisions about genetic engineering sometimes think more about the benefits to large corporations and industries and not enough about how the effects of genetic engineering might hurt humans and other animals.

Learning Set 4

Back to the Big Challenge

Make Recommendations About Developing a New Rice Plant That Will Produce More Rice and More Nutritious Rice

The Rice for a Better World Institute

Research Announcement

To: All Collaborating Scientists

From: The Rice for a Better World Institute (RBWI)

Subject: Research Update

The work you have done to learn more about genes and chromosomes should be very helpful in the last stage of finding a rice plant that can produce more rice or more nutritious rice. We recently heard about ways of selecting the genes in cells so that plants can have the traits we need. You now have enough information to help us decide what traits our new rice plant should have and the DNA sequences it needs on its chromosomes. We want you to do both these things for us. As you do that, remember that caterpillars have become a big problem in the field. Resistance to caterpillars is critical to growing more rice and more nutritious rice. So we will need to plant rice that can protect itself from caterpillars.

The table on the next page will help you identify traits the new rice plants should have. We have identified four varieties of rice that are close to what we need. Each has four traits. But none is perfect for our needs. One variety is resistant to caterpillars but has other traits that are not beneficial. Choose traits for the new rice by selecting traits from the four varieties of rice listed in the table.

Leaf blight disease

Rice blast disease

Rice variety	Traits	DNA sequence
A	Drought resistant high fiber caterpillar resistant weak stems (will not stand up in windy conditions)	AACG TACC GTGC CCCT
B	High starch requires less fertilizer maggot resistant high number of stem tumors (most plants die before they produce seeds)	ATAC GGAA ATTT ATGG
C	High levels of vitamins and minerals weak root system (cannot get enough nutrients) rice grains don't mature (are too small to eat) leaf blight resistant	GGGC CCGG AACC TTTG
D	High number of grains per plant rice grains have no germ (no starch) small leaves (cannot grow in cloudy conditions) high level of rice blast disease	TAGC CTTA AGCG CGGT

You will need to choose four traits for the new rice. One should be resistance to caterpillars. You can choose the other three. However, we have one more constraint we need you to consider. We will not be able to produce rice that has no defects. We can, however, choose which defect the rice will have.

Caterpillars have become a big problem for rice growing. It is important to the farmers that the rice they plant is resistant to being eaten by caterpillars. Your job is to select four rice traits that would make the best rice that is resistant to caterpillars while growing more rice or more nutritious rice. Some groups will make recommendations about a rice variety that produces more rice and is also caterpillar resistant. Some groups will recommend a rice variety that is more nutritious and resistant to caterpillars.

Identify Criteria and Constraints

Begin your discussion by focusing on the criteria and constraints you need to address for this part of the challenge. Record your list of criteria and constraints in a Criteria and Constraints table like the one on the facing page. Begin by identifying the criteria, the goals that must be satisfied. Then identify the constraints of the challenge, factors that limit your solution.

Criteria	Constraints

Plan Your Solution

Choose Your Rice's Traits

Work as a group to select four traits for a new rice plant. Examine the traits of all the varieties of rice in the table from the *RBWI*. Select traits from those listed in the table. As you examine the rice traits, notice that some of the traits are important for achieving the goal of developing a rice plant that produces more rice, and some are important for achieving the goal of producing more nutritious rice. For example, some of the traits may help the plant survive in difficult environmental conditions. Plants with traits that help them survive are likely to produce more rice than plants that do not have survival traits. Notice also that each rice has some genetic defect that you do not want in a new plant.

Think about these questions as you discuss which four traits the rice should have.

Rice Traits			4.1WC.2

Name: _____ Date: _____

Our Goal: _____
(more rice or more nutritious rice)

Environmental conditions we are taking into account and why:
_____
_____
_____

Rice Trait	Why did you choose this trait?	DNA Sequence	Matching Sequence
Caterpillar Resistance		G T G C	C A C G

- What environmental conditions should you take into account? Keep in mind that to pass their seeds to the next generation, plants must grow to maturity.

- Scientists will not be able to remove all defects from the new plants. Therefore, one constraint is that you will have to leave in one trait that would be a defect in your plant. Which one will you leave in? Why did you select that one?

- Which two beneficial traits did you decide to select for more rice or more nutritious rice? Why did you choose those traits?

GENETICS

Record on your *Rice Traits* page the environmental conditions you are taking into account and the rice traits you chose to include. Include the reasons you chose that trait. You will fill in the DNA sequence later.

Reflect

Answer the following questions about how you made your decisions.

1. Describe in detail how you determined the beneficial traits for your new rice.

2. How did you decide which defect your new plant should have?

Identify Your Rice's DNA Sequence

The traits you have identified will be coded in DNA as genes on your rice's chromosomes. The table shows the DNA base sequence associated with each rice trait. Add the gene sequence for each trait you have chosen to the column labeled "DNA Sequence" on your *Rice Traits* page. Then in the last column, add the matching sequence for each. The base pairing for Caterpillar resistance is already recorded as an example. Information about base pairings of DNA from the last section will help you fill in the pairings for each of your rice's traits.

Recommend

The traits you chose for the rice need to be described to the *RBWI* and the farmers. You need to present your trait choices and identify why you chose these traits rather than other ones. Develop recommendations for each of the three traits you chose for your rice. Use a *Create Your Explanation* page for each recommendation. Remember that your recommendation will be your claim. Add evidence and science knowledge that support each recommendation to the *Evidence* and *Science knowledge* portions of each *Create Your Explanation* page. Then develop a statement that uses the evidence and science knowledge to support your recommendation. Remember that a

---

Create Your Explanation

Name:_____ Date:_____

Use this page to explain the lesson of your recent investigations.

Write a brief summary of the results from your investigation. You will use this summary to help you write your Explanation.

Claim – a statement of what you understand or a conclusion that you have reached from an investigation or a set of investigations.

Evidence – data collected during investigations and trends in that data.

Science knowledge – knowledge about how things work. You may have learned this through reading, talking to an expert, discussion, or other experiences.

Write your Explanation using the **Claim**, **Evidence**, and **Science knowledge**.

---

good explanation will tell the *RBWI* scientists and farmers why your recommendation is a good one.

Think about starting your explanation statements with "*If*," "*When*," or "*Because*." For example, you might begin an explanation by writing, "If the farmers need to plant rice that contains more starch and is caterpillar resistant…" Even better would be a recommendation of the form, "Because the rice needs to contain more starch and caterpillar resistance…" Then complete the statement with your trait recommendation and the evidence or science knowledge that supports it.

Communicate Your Solution

Solution Briefing

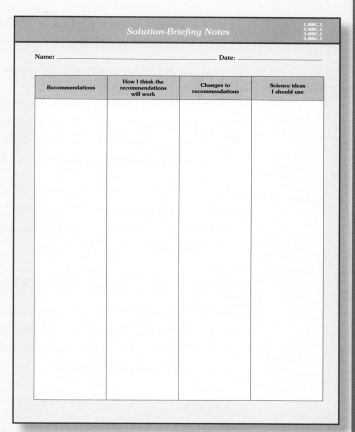

After you have developed your recommendations, you will communicate them to one another in a *Solution Briefing*. To prepare for this briefing, revisit the criteria and constraints you identified at the beginning of *Back to the Big Challenge*. Use the following questions to plan your briefing presentation:

- Which environmental factors did you take into account? How will the traits of your new rice address the environmental concerns?

- How did you choose a genetic defect for your new rice? How does it affect the "good" traits in your rice plant?

- Which traits did you decide were important for your new rice?

- How does your new rice meet the criteria?

- How did the constraints affect your recommendations?

- What information did you use to help you make your recommendations?

- What ideas did you think about along the way, and why did you not recommend them?

- What questions do you still have?

As you listen to the presentations of other groups, make sure you understand how they made decisions about the seed they would produce. If you do not understand something, or if they did not present something clearly enough, ask questions that help you understand why they chose their four traits.

As you listen, record your notes on a *Solution-Briefing Notes* page. Notice how your recommendations are similar to and different from those of other groups who worked with the same criterion. Be prepared to discuss the similarities and differences.

Revise Your Recommendations

As a class, compare the recommendations different groups made. Identify the new rice varieties that you think would survive best, produce the most rice, and produce the most nutritious rice. Decide if there is a combination of four traits that would achieve all of those criteria.

Update the *Project Board*

The last column of the *Project Board* helps you pull together everything you have learned in the Unit. This is the place to record the information from your readings and activities in this Unit that might help you address the *Big Challenge*—to make recommendations on developing a new rice plant that will produce more rice and more nutritious rice. Add recommendations about producing more rice or more nutritious rice that is resistant to caterpillars to the *Project Board* so you can return to it when you address the *Big Challenge* and answer the *Big Question* at the end of the Unit.

Address the Big Challenge

Final Recommendations

Make Recommendations About Developing a New Rice Plant That Will Produce More Rice and More Nutritious Rice

Now it is time to make the final recommendations and describe your thinking about this challenge to the class. Read the letter from the *RBWI*, and pay attention to the criteria and constraints listed there. After reading the letter and revising your criteria and constraints lists, you will combine everything that you found out in the Unit and do a final set of revisions on the recommendations you have made throughout the Unit. You will then present all your ideas to your class in a *Solution Showcase*. This presentation is the final project of this Unit and gives you and others a chance to share all the ideas.

Research Announcement

To: All collaborating scientists

From: The Rice for a Better World Institute (RBWI)

Subject: Research update

We thank you for all the help you have given us as we try to develop a rice plant that will produce more rice and more nutritious rice for farmers in the Philippines. We are grateful for all the work you have done for us and the farmers. We eagerly await your final recommendations.

As you develop your final recommendations, we want you to address this question:

Which rice varieties should the farmers plant?

As you make your decision we want you to think about these things:

1. The rice plant must be resistant to caterpillars so the farmers can reduce their use of pesticides. Pesticides are harmful to the farmers, their families, and the environment.

2. The farmers need their rice to grow even if the weather is very dry. The rice fields have to produce a good yield in both normal weather and in drought.

3. The rice must be as nutritious as possible. Rice is a staple food for people in the Philippines. Everyone there eats it and depends on it for nutrition.

4. Determine if you think it might be beneficial to grow different types of rice. If you think that would be a good recommendation, why would it be helpful to the nutrition or yield of rice plants?

We eagerly await your recommendations on how we should proceed in this challenge.

Revise Criteria and Constraints

Begin planning your final recommendation by identifying the full set of criteria and constraints. You can do this by revising criteria and constraints you have previously identified. Begin by revising criteria— the goals you should aim to achieve. Revise the constraints— the conditions that limit the way you can achieve the challenge. Revise a previous criteria and constraints page, or record the full list on a new criteria and constraints page.

Plan Your Solution

Pulling together what you have learned in this Unit and using that information to address the *Big Challenge* requires careful thought about what you have read and the recommendations you have made. Begin by thinking through these questions with your group. The questions will help you consider the *Big Challenge*. Discuss the answers and use your records, recommendations you made earlier, and information from the book and other resources to develop your ideas.

1. Traits —
   What traits do you think are the most important? Why do scientists need to consider the traits of plants when they think about using technology to improve yields and grow more nutritious crops?

2. Inheritance of traits —
   How are traits in plants inherited? How can you use this information to make sure that the traits of plants continue to be traits people need?

3. Environmental interaction —
   How can you take into consideration the environment in which a plant will grow? Why is this an important consideration? Why is it important to know how genes and the environment interact when you are developing a new plant in the laboratory?

4. Genetic engineering —
   Genetic engineering involves manipulating genes in the laboratory. How is this process done? How might you use it to develop new plants?

5. Benefits of genetically modified crops —
   Developing new plants in the laboratory has advantages. For one, it takes less time than using selective breeding in the field. How can genetically modified crops help people?

6. Dangers with genetically modified crops —
   Developing new plants in the laboratory and then planting them in fields can be dangerous to the environment. How can scientists control for these dangers?

7. Issues of agriculture —
   What resources are necessary for growing rice? How can these resources be managed? What are some issues with growing only one type of crop (monoculture) in a field or area? What might be some advantages of growing different types of rice? What are some problems with pesticides and fertilizers?

Revise Your Recommendations

After considering the answers to the questions above, revisit the recommendations you made earlier in the Unit. Your recommendations have been about the types of rice that are important to grow to meet the challenge of the Unit, the types of crosses you could make and the reasons for those crosses, and specific directions to farmers about how to plant and grow the rice. Decide what changes you need to make to your recommendations and work on revising them. Be sure your recommendations include directions to the farmers about how many varieties of rice you think the farmers should plant.

Complete a new *Create Your Explanation* page for each of your three recommendations. Be as precise as you can in making your recommendations, supporting them with evidence and science knowledge, and explaining why they should be trusted. You will probably need to refer to data from earlier investigations. The explanations and recommendations you and others made during the Unit might help you. Be sure to refer to the *Project Board* as a resource.

Communicate Your Recommendations

Solution Showcase

After all groups have made their final recommendations, everyone will complete this challenge by sharing recommendations in a *Solution Showcase*. The goal of a *Solution Showcase* is to help your class better understand how your group approached the challenge.

Your *Solution Showcase* poster and presentation should include:

- Your final recommendations on how the farmers or scientists should develop a new rice plant with the traits for the criterion you were addressing.

- How you accounted for the constraints and criteria of the challenge.

- How you took into account any special problems the methods you recommended might present.

You must include all of the possible problems you see in your plan. Solutions to problems can often cause other problems. Describe the problems you see in your plan, and suggest how your plan might minimize the problems.

Your presentation during a *Solution Showcase* should include the history of your plan. Review your original recommendations, and tell the class how you revised your plan and why you made your decisions. Make sure to present your reasons for the changes you made.

Your presentation should include detailed instructions to the scientists and farmers about how to address the challenge you were working on. Your instructions should include as much detail as possible. Be prepared to describe how the features of your plan will address the challenge. Be sure to give credit to others who helped you improve your plan.

During this Solution Showcase, you will see several solutions to the challenge. It will be important to provide support for your solution. Be sure to discuss how your final plan includes the explanations and recommendations that the class generated throughout the Unit. Explain why you think your solution is a good one.

As you listen to others present their recommendations, think about the following questions:

- How do the recommendations address the challenge of the Unit? What ideas of the Unit were left out?

- What else could the group have included in their recommendations or discussions to make its recommendations more complete?

- How did the group address the criteria and constraints of the challenge? Which criteria and constraints might the group have addressed better? How would you suggest group members improve their recommendations?

Answer the Big Question

How Can Knowledge of Genetics Help Feed the World?

The *Big Challenge* of this Unit focused on making recommendations about developing a rice plant that produces more rice and more nutritious rice. Developing a rice plant to help meet the goals of the challenge is one example of how knowledge of genetics can help feed the world. Growing rice that provides higher yields and better nutrition allows people to live healthier lives. Now you will use all the information from this Unit to answer the *Big Question: How can knowledge of genetics help feed the world?*

Throughout the Unit, you have read about ways that genetic knowledge has made significant changes in how people live. Sometimes these changes affect individuals, like the progress made in breast-cancer detection and the treatment of cystic fibrosis. Sometimes the focus is on changing entire populations of organisms, as in the rice challenge you have been working on. You also read about dog breeding and how specific breeds of dogs have been developed to meet the needs of people. Each of these examples may provide some part of the answer to the *Big Question.*

You will answer the *Big Question* in three steps. First, you will identify two possible ways that knowledge of genetics could be used to help feed the world. Second, you will identify potential advantages and disadvantages of the changes you propose. Then, you will develop a statement about how knowledge of genetics can be used to help feed the world. Finally, you will present your answers to the class in a *Solution Showcase.*

Plan Your Answer

Identify Two Ways Knowledge of Genetics Can Be Used to Help Feed The World

Begin to answer the question by thinking about possible ways knowledge of genetics could be used to help feed the world. Consider all the different foods people eat, and identify two ways genetics might improve those foods. There are several ways you can think about improving foods; you can select one of the ways listed below or think of your own.

- You might use your knowledge of genetics to improve the nutritional value of a food, as you did with rice. Make sure you select a food that is a staple, an important food many people eat. This will make the change you consider important to more people.

- One way to improve the foods people eat is to make shipping some foods easier. Changing food to make shipping easier, like developing strawberries that are firmer and last longer during shipping, could make these foods cheaper for people to purchase.

- Genetics could be used to make some important foods more plentiful. One way this might happen is to increase the amount of food that grows on a plant. Shortening the plants' life cycle or developing a rice plant with more rice grains per plant could make some foods more plentiful.

- Important foods could be developed that are less likely to be eaten or harmed by pests. For example, making rice resistant to caterpillars means that farmers will not have to use as many pesticides. This would also be better for the ecosystem in which the rice is being grown. The land, the water, and other organisms would be safer because fewer pesticides would be used.

Identify Advantages and Disadvantages of Your Proposed Changes

As you think about how the genetic changes you propose can help feed people, consider the positives and negatives of your changes. Look carefully at the genetic changes you propose and use evidence to determine what might happen as a result of these genetic changes. For example, you read about Bt-corn. Although Bt-corn solves many problems for farmers, some scientists are concerned about the effect this crop might have on the environment. Scientists have had to consider all the changes in the environment that might occur as a result of using Bt-corn.

For each of the changes you are proposing, answer the questions on the next page to help identify the possible effects.

1. Which organisms did you choose to change, and why?

2. What trait are you trying to improve in each of your organisms?

3. Why did you choose that trait?

4. How will changing that trait help feed the world?

5. What processes will farmers or scientists use to change each organism? What steps in these processes will be complicated?

6. All organisms depend on other organisms in the environment. In what ways might changing the traits of your organisms affect other organisms? Remember to consider how the changes might affect other organisms your new organism depends upon. Also consider how changes might affect other organisms that depend upon your organism.

7. You are considering changing organisms to help people. What might be some other long-term effects of changing the organism? Think about how growing your new organism might affect the ecosystem it is grown in. How might growing your new organism affect the land, water, and other resources in the environment?

Create Your Explanation

Name:_____ Date:_____

Use this page to explain the lesson of your recent investigations.

Write a brief summary of the results from your investigation. You will use this summary to help you write your Explanation.

Claim – a statement of what you understand or a conclusion that you have reached from an investigation or a set of investigations.

Evidence – data collected during investigations and trends in that data.

Science knowledge – knowledge about how things work. You may have learned this through reading, talking to an expert, discussion, or other experiences.

Write your Explanation using the **Claim**, **Evidence**, and **Science knowledge**.

Develop Your Statement

After you have identified the possible advantages and disadvantages of changing organisms to benefit people, develop a statement that answers the *Big Question: How can knowledge of genetics help feed the world?* Use a *Create Your Explanation* page to develop your answer.

Your claim should include two parts: A statement about how knowledge of genetics can help feed the world and a second statement about what scientists should be careful about as they consider making genetic changes. Your claim might have this form: "Genetics can be used to ..., but scientists should be careful about ..."

Use your experiences in developing a new rice plant and your answers to the questions on the previous pages as evidence for your claim. Use what you have read about genetics, about genetic engineering, and about ecosystems as science knowledge that support your claim. Make sure your evidence and science knowledge really support your claim. Write your claim so it matches your evidence and science knowledge.

Develop a statement that ties together your claim, your evidence, and your science knowledge. Your statement should convince others of how they might use knowledge of genetics to help feed the world, what they should be careful about when developing new organisms and why.

Communicate Your Answer

Solution Showcase

After you have developed your answer to the *Big Question*, you will share it in a *Solution Showcase*. Your presentation should include the following:

- your answer to the *Big Question* (your claim)

- the genetic changes you propose, the potential advantages of each, and the challenges and potential dangers of each

- any other evidence and science knowledge you used to develop your answer

- your statement

Each group may have a different answer to the *Big Question* because there are many different traits and organisms to consider. As you listen, pay attention to the evidence that groups use to support their answer. Think about how the ideas in the Unit helped answer the *Big Question*. If you think a group has not been convincing, ask it for more information or present evidence or science knowledge that suggests a different answer. Remember to be respectful, even when you do not agree with your classmates.

Reflect

You have heard a number of different answers to the *Big Question*. You have thought about the answers and the evidence each group used to answer the question. Use the following questions to discuss any differences of opinion:

1. Why did groups have different answers to the *Big Question*?

2. What made this a difficult question to answer?

3. When you disagreed with your classmates, you used evidence and science knowledge to help convince them of your answer. This is the way scientists discuss their ideas. When scientists disagree, they present their evidence to one another. Then they discuss how their evidence matches their answers. One way they begin to solve these disagreements is to think about gathering more evidence. To do this, they think of questions they need to investigate. What further questions might your class investigate to help you resolve any disagreements?

English & Spanish Glossary

A

adult leukemia Cancer of the white blood cells in adults.

leucemia en adultos Cáncer de las células blancas en adultos.

agriculture The production of food and other goods by growing plants and raising animals.

agricultura La producción de alimentos y otros bienes por medio del cultivo de plantas y crianza de animales.

alleles Different forms of a gene.

alelos Formas diferentes de un gen.

anaphase The step of mitosis during which the strands of the chromosomes are pulled apart by spindle fibers and move toward opposite ends of the cell.

anafase El paso de la mitosis durante el cual las hebras de un cromosoma son separadas por fibras del huso y movidas hacia los extremos opuestos de la célula.

anemia A low number of red blood cells, which carry oxygen in the blood to the body cells.

anemia Una cantidad baja de células o glóbulos rojos, los cuales transportan el oxígeno en la sangre hacia las células del cuerpo.

anther The structure on the stamen of flowers where pollen and sperm are produced.

antera La estructura en el estamen de las flores donde se producen el polen y la esperma.

asexual reproduction One parent cell divides into two cells, and each of the new cells has the same genetic material as the parent.

reproducción asexual Una célula madre o troncal se divide en dos células, y cada una de las células nuevas tiene el mismo material genético que la troncal.

B

bases Chemical structures that make up the rungs of a DNA ladder.

bases Estructuras químicas que constituyen los niveles de una escalera ADN.

binary fission The simplest form of asexual reproduction, in which the parent cell divides in two. Each cell has the same genetic material. Binary fission is used by protozoa, bacteria, and some algae.

fisión binaria La forma más simple de una reproducción asexual en la cual la célula madre se divide en dos. Cada célula tiene el mismo material genético. La fisión binaria es usada por el protozoa, las bacterias y algunas algas.

English & Spanish Glossary

biome Large land areas with similar environmental characteristics and similar types of plants.

bioma Areas grandes de terreno con características ambientales similares y tipos de plantas similares.

blending An equal mixing of traits.

fusionar Una mezcla igual de rasgos.

bran The skin of a grain.

salvado La piel de un grano.

budding A method of asexual reproduction in which an outgrowth forms on the parent and eventually breaks off and lives independently.

germinación Un método de reproducción asexual en la cual una excrecencia se forma en el progenitor y eventualmente se rompe y vive independientemente.

bug An insect with sucking mouthparts.

sabandija Un insecto con partes bucales para succionar.

C

carbohydrate A complex sugar. Carbohydrates provide energy when digested.

carbohidrato Una azúcar compleja. Los carbohidratos proveen energía cuando se digieren.

carpel The female part of flowers.

carpelo La parte femenina de las flores.

cell division The splitting of a parent cell into two daughter cells.

división celular El desdoblamiento de una célula madre en dos células hijas.

cell wall A rigid, protective covering in plant cells.

pared celular Una cubierta protectora y rígida en las células de las plantas.

centrioles Very small structures, found in animal cells, that produce the spindle fibers.

centriolos Estructuras muy pequeñas, encontradas en las células animales, que producen las fibras del huso.

cereal The edible seed of a grass plant; a grain.

cereal La semilla comestible de una hierba; el grano.

chest physiotherapy A treatment used for removing the thick mucus that forms in the lungs of a person with cysttic fibrosisf.

fisioterapia del pecho Un tratamiento utilizado para remover la mucosidad gruesa que se forma en los pulmones de una persona con fibrosis cística.

chromosomes Strands of genetic material inside the cell that contain the traits of an organism.

cromosomas Hebras de material genético dentro de una célula que contiene los rasgos de un organismo.

cloning An asexual reproductive process in which a group of genetically identical cells are produced by the division of a single cell.

clonación Un proceso de reproducción asexual en la cual un grupo de células idénticas genéticamente son producidas por medio de la división de una célula sencilla.

co-dominance Neither allele of a pair hides the other. Both alleles are dominant to the same extent; no blending occurs.

codominancia Ningún alelo de un par esconde el otro. Ambos alelos son dominantes en el mismo grado, no ocurre ninguna fusión.

constraints Factors that limit how you can solve a problem.

limitaciones Factores que limitan cómo puedes resolver un problema.

criteria Goals that must be satisfied to successfully achieve a challenge.

criterios Metas que se deben cumplir para alcanzar satisfactoriamente un reto.

English & Spanish Glossary

cross To breed two different varieties of plants to produce offspring with a mixture of traits

cruzar Aparear dos variedades diferentes de plantas para producir progenie con una mezcla de rasgos.

cross-pollination The transfer of pollen on one plant to the female part of another plant.

polinización cruzada La transferencia de polen en una planta a la parte femenina de otra planta.

cystic Fibrosis (CF) A hereditary disease that causes the body to produce thick, sticky mucus in the lungs, liver, pancreas, and intestines.

Fibrosis cística Una enfermedad hereditaria que ocasiona que el cuerpo produzca mucosidad pegajosa y gruesa.

cytoplasm A jelly-like substance that fills up the space inside the cell.

citoplasma Una sustancia gelatinosa que llena el espacio dentro de una célula.

D

dating Determining the age of sedimentary rocks or fossils.
datación Determinar la edad de las rocas sedimentarias o fósiles.

daughter cells The two cells that result from cell division
células hijas Las dos células resultantes de la división celular.

diversity An environment with many plants and animals that depend on each other for survival.
diversidad Un medio ambiente con muchas plantas y animales que dependen unos de los otros para sobrevivir.

DNA The genetic material in a cell that stores and transmits genetic information from one generation to the next.
ADN El material genético en una célula que almacena y transmite la información genética de una generación a la próxima.

domestic (domesticated) Bred or tamed for the use of people.
doméstico (domesticado) Criado o domado para el uso de las personas.

dominant The allele that masks the expression of the recessive allele.
dominante El alelo que oculta la expresión de un alelo recesivo.

donor organism An organism that supplies a gene for another organism.
organismo donante Un organismo que suple un gen para otro organismo.

double helix A twisted ladder structure.
doble hélice Una estructura en forma de escalera contorsionada.

Down syndrome A genetic disorder that results when chromosome number 21 fails to separate during meiosis.
síndrome de Down Un desorden genético que resulta cuando el cromosoma 21 falla en separarse durante la meiosis.

drought A long period, lasting weeks or months, with little or no rainfall.
sequía Un período largo, que dura semanas o meses, con poca o ninguna lluvia.

E

egg The female sex cell; the reproductive cell that contains the female's genetic information; also called the female gamete.
huevo La célula sexual femenina; la célula reproductiva que contiene la información genética femenina; conocido también como el gameto femenino.

English & Spanish Glossary

endosperm Nourishment that surrounds the germ (embryo) of a seed.
endospermo Nutrientes que envuelven el germen (embrión) de una semilla.

environmentalist Scientists who study the environment, and other people who are concerned about the environment.
ambientalista/ecologista Científicos que estudian el medio ambiente, y otras personas que están preocupadas por el medio ambiente.

erosion The loss of soil through the actions of water and wind.
erosión La pérdida de terreno por medio de las acciones del agua y el viento.

evolution The change in the frequencies of genes in a population over time.
evolución El cambio en las frecuencias de los genes en una población a través del tiempo.

expressed (to express) Shown (to show).
expresado (expresar) Mostrado (mostrar).

F

F1 generation The first generation of offspring from a breeding.
generación F1 La primera generación de progenie producto de un apareamieno.

Project-Based Inquiry Science

F₁ The first generation of hybrids from a breeding.

F₁ La primera generación de híbridos producto de un apareamiento.

F₂ generation The second generation of offspring from a breeding.

generación F₂ La segunda generación de progenie producto de un apareamiento.

F₂ The second generation of hybrids from a breeding.

F₂ La segunda generación de híbridos producto de un apareamiento.

fertilization The fusion of the cell that contains the male traits with the cell that contains the female traits.

fertilización La fusión de las células que contienen los rasgos masculinos con la célula que contiene los rasgos femeninos.

fossil record All the fossils ever found.

récord fósil Todos los fósiles que se han encontrado hasta ahora.

fossils The remains of living organisms preserved in rocks in Earth's crust.

fósiles Los restos de organismos vivos conservados en rocas en la corteza terrestre.

fragmentation An asexual reproductive process in which an organism breaks into pieces, and each piece grows into a new individual.

fragmentación Un proceso reproductivo asexual en el cual un organismo se rompe en pedazos, y cada pedazo crece en un nuevo individuo.

G

gene The location on the chromosome that contains the instructions for a particular trait.

gen El lugar en el cromosoma que contiene las instrucciones para un rasgo particular.

genetic engineering Changing the genes of organisms.

ingeniería genética El cambio de los genes de los organismos.

genetic engineers Scientists who change the genes of organisms.

ingenieros genéticos Científicos que cambian los genes de los organismos.

genetic material Genetic information in an organism that is passed down from generation to generation.

material genético La información genética en un organismo que es pasada de generación en generación.

English & Spanish Glossary

genetically modified Used to describe food or organisms that have been genetically engineered by inserting a gene from another organism.

modificado genéticamente Utilizado para describir alimentos u organismos que han sido creados genéticamente insertando un gen de otro organismo.

genetics The science of how characteristics are passed down from one generation to the next.

genética La ciencia que estudia cómo las características pasan de una generación a otra.

genome The complete set of genes of an organism.

genoma El conjunto completo de genes de un organismo.

genotype The genetic makeup of an organism.

genotipo La composición genética de un organismo.

geologist A scientist who studies the origin, history, and structure of Earth.

geólogo Un científico que estudia el origen, historia, y la estructura de la Tierra.

germ (embryo) The part of the seed from which a new plants grows.
germen (embrión) La parte de la semilla de la cuál crecen plantas nuevas.

grain Usually a type of grass grown for its edible seeds. Grains include wheat, rice, corn, oats, barley, buckwheat, quinoa, millet, and others. Also used to describe the seed of the grain in plants, as in rice grain.
grano/cereal Generalmente un tipo de hierba que se cultiva por sus semillas comestibles. Los granos incluyen trigo, arroz, maíz, avena, cebada, alforfón o trigo negro, quinoa, mijo o millo y otros. Es utilizado también para describir la semilla de los granos en las plantas, como en el grano del arroz.

H

half-life The time it takes for half of an amount of a radioactive element to change into a new stable element.
vida media El tiempo que le toma a la mitad de una cantidad de un elemento radioactivo para cambiar a un elemento estable nuevo.

hectare A metric unit of area measurement equal to 2.471 acres.
hectárea Una unidad métrica de área midiendo igual a 2.471 acres.

heredity or **inheritance** The passing along of information from one generation to the next.
herencia El traspaso de información de una generación a la próxima.

heterozygous An organism that has two different alleles for the same gene.
heterócigo Un organismo que tiene dos alelos diferentes para el mismo gen.

homozygous An organism that has two identical alleles for a particular gene.
homocigoso Un organismo que tiene dos alelos idénticos para un gen particular.

hull (husk) The tough, outer layer on a seed.
vaina (cáscara) La capa externa y dura en una semilla.

Huntington's disease A fatal genetic disease caused by a dominant allele, that affects the nervous system.
enfermedad de Huntington Una enfermedad genética fatal causada por un alelo dominante, que afecta el sistema nervioso.

hybrid The offspring of two different types of the same organism.
híbrido La progenie de dos tipos diferentes de un mismo organismo.

GENETICS

English & Spanish Glossary

hypothesis (plural: hypotheses) A prediction of what will happen between an independent (manipulated) variable and a dependent (responding) variable.
hipótesis Una predicción de lo que ocurrirá entre una variable independiente (manipulada) y una variable dependiente (de respuesta).

I

incomplete dominance One allele of a pair cannot completely hide the traits of its partner; this results in a "blending" of traits in the first generation of offspring.
dominancia incompleta Un alelo de un par no puede esconder completamente los rasgos de su pareja; esto resulta en una "mezcla" de rasgos en la primera generación de la progenie.

inherit Receive traits from the previous generation.
herencia Los rasgos recibidos de la generación anterior.

inheritance The passing of traits from one generation to the next.
herencia El paso de rasgos de una generación a la próxima.

insecticide A substance used to kill insects.
insecticida Una sustancia utilizada para matar insectos.

interphase The step, before mitosis begins, during which a cell prepares to divide.
interfase El paso, antes de comenzar la mitosis, durante el cual una célula se prepara para dividirse.

J

jaundice A condition arising when pigments from the gall bladder invade the blood. The skin and eyes become yellow.

ictericia Una condición resultante cuando pigmentos de la besícula biliar invaden la sangre. La piel y los ojos se tornan amarillos.

L

larvae (singular: larva) The first stage of an insect after it leaves the egg.

larva La primera etapa de un insecto después de dejar el huevo.

M

maize Another name for domesticated corn.

maíz Otro nombre para el maíz doméstico.

malnutrition A condition resulting from not enough food or lack of the proper food.

desnutrición Una condición que resulta de la falta o escasez del alimento apropiado.

mapping (genes) Finding the location of every gene on a chromosome.

cartografía genética (genes) Encontrar la localización de cada gen en un cromosoma.

meiosis Cell division that produces sex cells with one half the number of chromosomes found in each body cell.

meiosis División celular que produce gametos con una mitad de la cantidad de cromosomas encontrados en cada célula del cuerpo.

membrane A covering around a cell that allows materials to flow in and out.

membrana La cubierta alrededor de una célula que permite que los materiales fluyan hacia adentro y hacia afuera.

metaphase The step of mitosis where the chromosomes line up across the center of the cell.

metafase El paso de la mitosis donde los cromosomas se alinean a lo largo del centro de la célula.

mitosis The duplication and splitting apart of chromosomes during cellular division.

mitosis La duplicación y separación de los cromosomas durante la división celular.

monoculture Using the land for the growing of only one crop.

monocultivo El uso del terreno para el cultivo de una cosecha solamente.

mucus A secretion of the body.

mucosidad Una secreción del cuerpo.

English & Spanish Glossary

mutation Changes to the genetic material of an organism.
mutación Cambios al material genético de un organismo.

N

native plant From the local area and best adapted to the local climate.
nativa plantas Perteneciente al área local y que mejor se adapta al clima local.

natural selection The differences in survival and reproduction among members of a population as a result of selection pressure.
selección natural Las diferencias de supervivencia y reproducción entre los miembros de una población como resultado de la presión de selección.

naturalist A person who studies the plants, animals, and environment of an area.
naturalista Una persona que estudia las plantas, los animales, y el medio ambiente de un área.

nuclein The name Friedrich Miescher gave to the material in cells that we now call DNA.
nucleíno El nombre dado por Friedrich Miescher al material en las células que ahora conocemos como ADN.

nucleus The control center of the cell. It contains DNA.
núcleo El centro de control de la célula. Contiene ADN.

Project-Based Inquiry Science

O

offspring The descendants of a person, animal, or plant.

progenie Los descendientes de una persona, animal o planta.

organelles Structures in the cytoplasm that perform special functions in the cell.

orgánulos Estructuras en el citoplasma que llevan a cabo funciones especiales en la célula.

ovary The part of the plant that makes the female cells containing the traits from the female parent.

ovario La parte de la planta que compone las células femeninas que contienen los rasgos de la madre.

ovule (egg cell) A tiny, egg-like structure in flowering plants that contains the female traits and develops into a seed after fertilization.

óvulo Una estructura pequeña, en forma de huevo en las plantas con flores que contiene los rasgos femeninos y se desarrolla en una semilla después de la fertilización.

P

p generation The parental generation in a breeding.

generación p La generación parental en un apareamiento.

paleontologists Scientists who study fossils to learn about living things that existed in the past.

paleontólogos Científicos que estudian los fósiles para aprender sobre los organismos vivientes que existieron en el pasado.

parent cell The original cell before it divides.

célula madre La célula original antes de dividirse.

petal A flower's outer protective covering, usually colored. Used also to attract insects and animals for pollination.

pétalo La cubierta protectora externa de una flor, usualmente con color. Utilizada también para atraer insectos y animales para la polinización.

phase A step or a stage in the process of development.

fase Un paso o etapa en el proceso de desarrollo.

phenotype The physical characteristics of an organism.

fenotipo Las características físicas de un organismo.

English & Spanish Glossary

pistil The female reproductive organ of a flower; may be made up of a single carpel or of two or more fused carpels.

pistilo El órgano reproductivo femenino de una flor; se puede componer de un carpelo sencillo o dos o más carpelos fundidos.

pollen A structure in flowering plants that has cells that contain traits of the male parent.

polen Una estructura en las plantas florales que tiene células que contienen rasgos del padre.

pollination The delivery of pollen to the female part of the plant.

polinización La transportación de polen a la parte femenina de la planta.

prairie Large area of grasslands usually located in the interior of continents, a biome with open fields and deep-rooted grasses.

pradera Zona de pastoreo extensa localizada usualmente en el interior de los continentes, un bioma con campos abiertos y hierbas con raíces profundas.

probability The chance that something will happen.

probabilidad la posibilidad de que algo suceda.

prophase the first step of mitosis during which the genetic material condenses into chromosomes. Each chromosome consists of two identical strands.

profase El primer paso de la mitosis durante el cual el material genético se condensa en cromosomas. Cada cromosoma consiste de dos hebras idénticas.

Punnett square A tool scientists use to investigate the possible combinations of genetic crosses.

cuadro o tabla de Punnett Una herramienta que los científicos utilizan para investigar las combinaciones posibles de los cruces genéticos.

R

radioactive dating A method of dating fossils by measuring the amount of a radioactive element in the fossils and in the rocks in which the fossils are found.

datación radioactiva Un método de datar los fósiles midiendo la cantidad de un elemento radioactivo en los fósiles y en las rocas en las cuales se encuentran los fósiles.

receptacle The main stem of a flower.

receptáculo El tallo principal de una flor.

recessive The allele whose expression is masked by the dominant allele.

recesivo El alelo cuya expresión es ocultada por un alelo dominante.

relative dating Determining which fossil is older than another by comparing the relative positions of the rock layers in which they are found.

datación relativa La determinación de cuál fósil es más antiguo que otro comparando las posiciones relativas de las capas de roca en las cuáles se encuentran.

reproduce sexually (sexual reproduction) One trait from each pair comes from the female parent, and one trait from each pair comes from the male parent.

reproducido sexualmente (reproducción sexual) Un rasgo de cada par proviene del pariente femenino, y un rasgo de cada par proviene del pariente masculino.

S

samples A piece or part taken from the whole group, whose properties are studied to gain information about the whole group.

muestras Una pieza o parte tomada del grupo completo, cuyas propiedades son estudiadas para obtener información sobre el grupo completo.

English & Spanish Glossary

sampling The process of selecting a suitable sample, or representative part,of the whole group.

muestreo El proceso de selección de una muestra adecuada, o parte representativa de todo el grupo.

sediment Small, solid pieces of material that come from rocks or living organisms.

sedimento Piezas sólidas y pequeñas de material que proviene de las rocas o de organismos vivos.

sedimentary rocks Rocks formed from the compression and cementing together of layers of sediment deposited in oceans, lakes, and swamps.

rocas sedimentarias Rocas formadas por la compresión y consolidación de capas de sedimento depositadas en los océanos, lagos y pantanos.

self-pollination The transfer of pollen on one plant to the female part of the same plant.

autopolinización La transferencia de polen en una planta a la parte femenina de la misma planta.

sepal A flower's outer protective covering, usually green.
sépalo La cubierta externa protectiva de la flor, usualmente de color verde.

sex cells In many organisms, the sex cells are egg cells and sperm cells. Both the egg and sperm are single cells. In humans, the egg is the largest single cell in the body, and the sperm is the smallest.
gametos En muchos organismos, los gametos (células sexuales) son óvulos y espermatozoides. Ambos, el óvulo y el espermatozoide, son células sencillas. En los humanos, el óvulo es la célula sencilla más grande y el espermatozoide la célula sencilla más pequeña.

sex chromosomes The chromosomes that determine the sex of an individual. In humans, these chromosomes are known as X and Y.
cromosomas sexuales Los cromosomas que determinan el sexo de un individuo. Estos cromosomas son conocidos como X y Y en los humanos.

sexual reproduction Reproduction that involves two parents.
reproducción sexual La reproducción que envuelve a ambos padres.

sickle-cell anemia A genetic disease, carried by a recessive allele, that affects the ability of the blood cells to carry oxygen.
anemia drepanocítica Una enfermedad genética, transportada por un alelo recesivo, que afecta la habilidad de las células de la sangre para transportar oxígeno.

species A group of organisms whose members have the same structural traits and can breed with one another.
especies Un grupo de organismos cuyos miembros poseen las mismas características estructurales y pueden aparearse unos con los otros.

sperm cells Structures that contain the male chromosomes.
espermatozoides Estructuras que contienen los cromosomas masculinos.

sperm The male sex cell; the reproductive cell that carries the male's genetic information to the female's egg, also called the male gamete.
esperma La célula sexual masculina. La célula reproductiva portadora de la información genética masculina al huevo femenino, conocida también como el gameto masculino.

English & Spanish Glossary

spindle A structure made up of tiny tubes that attach to the duplicated chromosomes and pull them apart in anaphase of mitosis.

eje celular Una estructura compuesta por tubos diminutos que se pegan a los cromosomas duplicados y los halan separándolos en la anafase de la mitosis.

spore formation An asexual reproductive process in which an organism forms a special cell called a spore.

formación de esporas Un proceso reproductivo asexual en el cual un organismo forma una célula especial conocida como espora.

stamen The male part of the flower.

estamen La parte masculina de la flor.

staple food A basic or necessary food item.

alimento básico Un artículo de alimento fundamental o necesario.

stigma The top part of the carpel where the pollen is deposited.

estigma La parte superior del carpelo donde se deposita el polen.

style The slender tube like part of the carpel.

estilo La parte delgada en forma de tubo del carpelo.

T

telephase The step of mitosis when the single strands of chromosomes gather at opposite ends of the cell and no longer become visible as separate chromosomes.

telefase El paso de la mitosis cuando las hebras sencillas de cromosomas se juntan en lados opuestos de la célula y ya no son visibles como cromosomas separados.

teosinte The wild ancestor of corn.

teosinte El ancestro silvestre del maíz.

trait A physical or behavioral characteristics of an individual that can be passed down to the next generation.

rasgos una característica físicas o de comportamiento de un individuo que pueden ser pasadas a la próxima generación.

transmission electron microscope A microscope that uses electrons instead of light to produce images of very small objects.

microscopio de transmisión electrónica Un microscopio que utiliza electrones en vez de luz para producir imágenes de objetos muy pequeños.

true-breeding Organisms that always pass their traits on to the next generation.

genéticamente puros Organismos que siempre pasan sus rasgos a la próxima generación.

tumor An abnormal growth of body tissue.

tumor Un crecimiento anormal del tejido corporal.

V

variation What makes one kind different from others of the same kind.

variación Lo que hace a una especie diferente de otras de la misma especie.

vegetative reproduction An asexual reproductive process in plants in which new cells separate from the parent and form new organisms.

reproducción vegetativa Un proceso reproductivo asexual en las plantas en el cual células nuevas se separan de la madre y forman organismos nuevos.

W

white blood cells Blood cells that protect the body against diseases.

células blancas Células en la sangre que protegen el cuerpo contra las enfermedades.

Z

zygote A fertilized egg.

zigoto Un huevo fertilizado.

Index

Index

Index

Index

Project-Based Inquiry Science

inheriting traits

See also inheritance

defined, GEN 49

insecticides

and Bt-corn, GEN 216, GEN 217

defined, GEN 8

as environmental issue, GEN 228

insulin, genetic engineering of, GEN 214

interphase

defined, GEN 179

in meiosis, GEN 194-GEN 195

in mitosis, GEN 179-GEN 183

J

jaundice, defined, GEN 85

L

larvae

of corn borer, GEN 215-GEN 215

defined, 216

length, as rice trait, GEN 29-GEN 31

light, required for growing rice, GEN 18

line graphs, GEN 106-GEN 111

defined, GEN 106

M

maize

artificial selection and, GEN 143-GEN 144

defined, GEN 143

malnutrition, defined, GEN 33

mapping (genes)

defined, GEN 207

Human Genome Project, GEN 207-GEN 211

medications, genetic engineering of, GEN 214

meiosis, GEN 191-GEN 195

defined, GEN 191

membrane, defined, GEN 202

Mendel, Gregor, father of modern genetics, GEN 49-GEN 53

Mendel's experiments, on inheritance, GEN 49-GEN 53, GEN 54-GEN 57

metaphase

defined, GEN 180

in meiosis, GEN 194-GEN 195

in mitosis, GEN 179-GEN 183

microscope

structure and use, GEN 172-GEN 174

transmission electron microscope, GEN 204-GEN 206

Miescher, Friedrich, "nuclein" discoverer, GEN 203

minerals, rice as source, GEN 32

mitosis, GEN 179-GEN 183

defined, GEN 179

modeling, GEN 76

importance in DNA study, GEN 205-GEN 206

Monarch butterflies, and Bt-corn, GEN 216-GEN 217

monocultures, GEN 137-GEN 140, GEN 143

defined, GEN 137

mucus, GEN 83

defined, GEN 82

mutation

in cystic fibrosis, GEN 84

defined, GEN 85

N

native (plants), GEN 138, GEN 139

defined, GEN 139

naturalists

Darwin and Wallace, GEN 114-GEN 117

defined, GEN 114

Index

P

paleontologists

defined, GEN 121

fossil examination, GEN 120-GEN 123

parameters

of Bug Hunt Speed simulation, GEN 106-GEN 111

defined, GEN 106

parent cells, GEN 179-GEN 183

defined, GEN 179

Pauling, Linus, molecular structure studies of, GEN 204-GEN 206

peas *See* garden peas

pest resistance, GEN 214

See also insecticides

in corn, GEN 215-GEN 216

in rice, GEN 166-GEN 167, GEN 178, GEN 219, GEN 226

petals, defined, GEN 46

P generation, GEN 62-GEN 64

defined, GEN 62

phases

defined, GEN 179

of mitosis, GEN 179-GEN 183

phenotypes

See also expression; genotypes; Reeze-ot models

defined, GEN 55

and dominant/recessive inheritance, GEN 55-GEN 57

field experiment results, GEN 73-GEN 75

and new hybrids, GEN 68-GEN 70, GEN 71-GEN 75

probability and, GEN 61-GEN 65

Philippines, importance of rice in, GEN 7-GEN 9, GEN 20

photosynthesis, defined, GEN 32

pistil, GEN 45-GEN 48

defined, GEN 46

plant populations

See also populations

adaptation in, GEN 103-GEN 105, GEN 113

artificial selection in, GEN 143-GEN 144

plants *See* flowering plants

pollen, GEN 45-GEN 48

defined, GEN 45

pollination, GEN 45-GEN 48

artificial cross-pollination, GEN 50-GEN 52

defined, GEN 45

types of, GEN 47

populations

See also animal populations; plant populations

adaptations in, GEN 103-GEN 113

artificial selection and, GEN 141-GEN 144, GEN 158

evolution and, GEN 118

extinction of, GEN 100

natural selection and, GEN 116

selection pressure and, GEN 95-GEN 102

prairies

defined, GEN 137

diversity and, GEN 137-GEN 139

predators

defined, GEN 103

selection pressure from, GEN 105-GEN 112

prey

See also predators

defined, GEN 103

probability, GEN 61-GEN 65

defined, GEN 61

Project Board, defined, GEN 11-GEN 12

Index

Index

Index

84 Business Park Drive, Armonk, NY 10504
Phone (914) 273-2233 Fax (914) 273-2227
www.its-about-time.com

Publishing Team

President

Tom Laster

Director of Product Development

Barbara Zahm, Ph.D.

Managing Editor

Maureen Grassi

Project Development Editor

Ruta Demery

Project Manager

Sarah V. Gruber

Assistant Editors, Student Edition

Gail Foreman

Susan Gibian

Nomi Schwartz

Assistant Editors, Teacher's Planning Guide

Danielle Bouchat-Friedman

Kelly Crowley

Edward Denecke

Heide M. Doss

Jake Gillis

Rhonda Gordon

Creative Director

John Nordland

Production/Studio Manager

Robert Schwalb

Production

Sean Campbell

Illustration

Dennis Falcon

Technical Art/ Photo Research

Sean Campbell

Michael Hortens

Marie Killoran

Equipment Kit Developers

Dana Turner

Joseph DeMarco

Safety and Content Reviewers

Edward Robeck

Barbara Speziale

Picture Credits

Picture Credits

Picture Credits